PHYSICS
FOR
FUTURE
PRESIDENTS

PHYSICS
FOR
FUTURE
PRESIDENTS

THE SCIENCE
BEHIND THE HEADLINES

RICHARD A.
MULLER

W·W·Norton & Company
New York
London

For information about permission to reproduce selections from this book,
write to Permissions, W. W. Norton & Company, Inc.,
500 Fifth Avenue, New York, NY 10110

For information about special discounts for bulk purchases,
please contact W. W. Norton Special Sales at
specialsales@wwnorton.com or 800-233-4830

Book design by Margaret M. Wagner
Production manager: Andrew Marasia

Library of Congress Cataloging-in-Publication Data

Muller, R. (Richard)
Physics for future presidents : the science behind the headlines / Richard A. Muller.
p. cm.
Includes bibliographical references and index.
ISBN 978-0-393-06627-2 (hardcover)
1. Physics—Social aspects. 2. Science and state. 3. Science in mass media. I. Title.
QC28.M85   2008
530—dc22
2008006075

W. W. Norton & Company, Inc.
500 Fifth Avenue, New York, N.Y. 10110
www.wwnorton.com

W. W. Norton & Company Ltd.
Castle House, 75/76 Wells Street, London W1T 3QT

To Luis Alvarez, *for teaching me
how to pursue scientific discovery,*
to Dick Garwin, *for showing me how to
be effective in national security,*
and to Art Rosenfeld, *for inspiring
my interest in energy and conservation*

CONTENTS

V · GLOBAL WARMING

Introduction

Are you intimidated by physics? Are you mystified by global warming, spy satellites, ICBMs, ABMs, fission, and fusion? Do you think all nukes, those in bombs and those in power plants, are basically the same? Are you perplexed by claims that we are running out of fossil fuels when there are counterclaims that we are not? Are you confused by the ongoing debate over global warming, when some prestigious scientists say that the debate is over? Are you baffled, bewildered, and befuddled by physics and high technology?

If so, then you are not ready to be a world leader. World leaders *must* understand these issues. The moment when you are being told that a terrorist left a dirty bomb hidden in midtown Manhattan is not a good time to have to telephone your local science advisor to find out how bad the situation really is. Nor is it a good time simply to assume the worst, to decide that all government resources must now be pulled off other projects to address this new emergency. You have to know enough to act wisely, quickly, proportionately.

Maybe you did study physics, enjoyed it, maybe even majored in it, and yet even now, after you got your degree, you *still* don't know the important difference between a uranium bomb and a plutonium bomb, or between ozone depletion and greenhouse warming. And when your friends ask you about spy satellites, you tell them what you read in the newspapers—because such details were never covered in your courses.

Many, if not most, important decisions today have a high-tech component. How can you lead your country into a clean-energy future if you don't understand solar power or how coal could be converted into gasoline? How can you decide important issues about research funding, arms control treaties, threats from North Korea or Iran, spying, and surveillance, if you understand only the political issues and not the technical ones? Even if you don't plan to be a world leader, how can you vote intelligently without understanding these issues?

Equally important to understanding the physics of modern life is unlearning the things that you may think are true but aren't. Mark Twain is often quoted as saying,

> *The trouble with most folks isn't their ignorance. It's knowin' so many things that ain't so.*

Ironically, this quote isn't even from Twain—as if to illustrate the aphorism itself. The quote is correctly attributed to Josh Billings, a nineteenth-century humorist.

Don't know the physics you need to know? Fortunately, you have found the solution, or at least the beginning of the solution. This book covers advanced physics, the stuff that world leaders need to know. I skip the math because you don't have time (or possibly the inclination) to master it. I move right to the important issues. When you understand the underlying principles, the physics, you need never again be intimidated by high tech. And if you ever need a detailed computation, you can always simply hire a physicist.

If you hate physics, it isn't your fault; you probably had a bad teacher. Some physicists intimidate outsiders by obscuring their knowledge in a fog of math. Among themselves, they are careful to distinguish physics from math. Graduate students are admonished, "Think physics—not math." But whenever it is a little tricky to explain something, the physics teacher has an easy way out: "Here, let me show you an equation that you can't possibly understand, and then you won't blame my poor teaching."

I offer only one caution. This book is not long enough to be a comprehensive review of anything. That means I leave out a lot about terrorism and counterterrorism, about energy and satellites and global warming. I don't apologize for that. Future presidents have additional things they need to know besides physics. Many people would like their president to be an expert in technology. I am more realistic. I try to cover in this book only the most essential facts and ideas, the key concepts that will help a president make better decisions. I could write a much longer book, but would a future president have time to read and study it?

I begin with terrorism because of its immediacy and because it draws us right into the subject of energy. After all, what is a bomb other than a quick way of delivering a huge amount of energy to a small space? What happened at the World Trade Center is much clearer when you understand the physics. Even biological weapons are easier to understand when you understand the relevant physics.

The second subject I cover is energy, a topic of central concern for all world leaders. Economies are made and broken by energy; countries go to war over energy; wasteful use of energy could destroy our ecology. Where do we get it, how do we use it, and can we avoid wasting it? There are many surprises about energy—facts that define and limit future technologies—everything from the possible hydrogen economy to the future of coal and oil and solar. Our love affair with fossil fuel, it turns out, has a foundation in physics.

Closely related to energy are nukes: radioactivity, bombs, reactors, dangers of waste—and the subject of the third major section of the book. A president must protect the public, but does that mean using nukes or ruling them out? There is no easy answer here, and decisions must be made on the details—physics details.

Our future clearly lies in space—or does it? It is thirty-nine years after Apollo 11, and we have not yet returned to the Moon. What's going on? What's holding us back? Is it economics, human will, or physics? What is the value of space? What does it mean to go into orbit? What are the real limitations? The fourth section of the book is devoted to space, satellites, and understanding the potential and limits of gravity.

Finally I touch on the hottest topic of all—global warming—a subject that spans so many different areas of physics that it has to have a section to itself. Moreover, it is a subject in which misinformation is as prevalent as truth. You may have to unlearn some cherished facts, but you need to do that to be able to make wiser decisions when you are president.

All of this is physics by total immersion. Don't pause too long if you find something confusing. Learn it like a foreign language: dive in, swim through it, wallow in it, and—most of all—enjoy it. Some material will be covered in more than one section. That's because the world is intertwined; nukes are relevant to terrorism, energy, space, and global warming. I regret if your favorite topic isn't here, but this book had room for only the topics of the most urgent interest. (The subject of UFOs, for instance, had to be left to a sequel.) I try hard to stay away from issues in politics, business, and diplomacy. Economics does creep in from time to time, but that is not the emphasis. This is the *physics* that you need to know to be a world leader. The rest is up to you.

I

TERRORISM

WHAT do you *think the next terrorist attack will be? A nuclear explosion? A dirty bomb? A dozen commercial airplanes blown up simultaneously? Or maybe something conventional—such as an airplane crashed into a skyscraper?*

If you are president, then anticipating and preventing such terrorism will be one of your top responsibilities. Sure, you'll get help—from your national security advisor, the CIA, the Department of Energy, your cabinet, and several hundred newspaper editorial writers. But the buck stops with you. If you fail to make the right decisions, you'll never forgive yourself, and neither will history. The responsibility is awesome. Are you sure you want the job?

It's not easy being president. But knowing some physics can help. Not all possible attacks are equally feasible or equally dangerous or equally difficult to counter. We all know the history of 9/11. Let's look at the events again, but this time from a physics perspective to help elucidate what happened and give us clues about the true dangers of the future.

1

NINE-ELEVEN

THE NATURE of the terrorist attack on the eleventh of September 2001 was unanticipated. Al-Qaeda used a weapon that released the energy of about 1.8 kilotons of TNT, substantially greater than the energy of the North Korean nuclear weapon test of October 9, 2006.

It wasn't the impact of the airplane that did the damage. The plane weighed 131 tons and was moving at 600 miles per hour. We can calculate the energy of motion, the *kinetic energy*, from the physics equation $E = \frac{1}{2}mv^2$. To use this equation, we have to use the right units[1] (that's often the hardest part); when we do that, we find that this energy was only 1 ton of TNT equivalent—1,800 times smaller than what the terrorists actually used. The kinetic energy was not what destroyed the buildings. Indeed, when hit by the airplanes, the towers of the World Trade Center hardly shook. Watch the film again (if you can stand it), but this time concentrate on the top of the building, above where the airplane strikes. Note that the upper part of the building barely moves. The impact itself had little impact.

The true source of the destructive energy was surprisingly simple: the 60 tons of jet fuel that each plane carried to propel it across the United States. Here is the surprising physics fact behind the attack:

One ton of jet fuel or gasoline, when burned in air, releases the energy of 15 tons of TNT. So 60 tons of gasoline releases the energy of 900 tons of TNT. For the two planes, the total was 1800 tons, 1.8 kilotons of TNT equivalent.

Gasoline has more energy than TNT? Yes, much more. In fact, even chocolate chip cookies have more energy than TNT. If you want to destroy a car, you can use a stick of TNT. But an equal weight of chocolate chip cookies, fed (for example) to teenagers with sledgehammers, can accomplish far more destruction. The chocolate chip cookies deliver about 5 food calories of energy per gram (the weight of about 2 chocolate chips), a number you can find in any diet book, whereas TNT delivers only 0.65 food calories per gram—nine times less.

That fact surprises most people, but it makes sense if you think about it. TNT is used not because of its high energy content, but because it can release its energy very rapidly. It can do this because (unlike gasoline or chocolate chip cookies) it doesn't have to be combined with air. The atoms in the TNT molecules are analogous to compressed springs held by a catch; release the catch, and the energy comes flying out. Break one molecule of TNT, and the energy breaks the adjacent catches. TNT then detonates as a chemical chain reaction. In a millionth of a second, the energy of the springs is transformed into energy of motion. The molecules have high velocity, and that means they are hot.

There are many ways to measure energy. In nuclear weapons treaties, the standard unit is *tons of TNT equivalent*. As defined by arms controllers, a ton of TNT is the energy of a million food calories (I'm not kidding), even though real TNT delivers only about two-thirds of this. The physicist's favorite unit is not the food calorie but the joule; one food calorie contains about 4200 joules.

The energy content of various materials is a key determinant, not only in terrorism but in many benign applications. A high-quality computer battery, for example, delivers only 1% of the energy of an equal weight of gasoline. That low number is the fundamental physics reason why most of us are not yet driving electric automobiles. We'll talk more about batteries in Part II, which focuses on energy. For now, let's delve deeper into the physics of the 9/11 attack.

The huge amount of energy in gasoline (and its other forms, including oil and jet fuel) has long made it an ideal substance to use in weapons. The history of such use probably dates back to the Byzantine Greeks and may be the secret of "Greek fire" (look it up in Wikipedia). Gasoline was the key component of the Molotov cocktails used in Spain in the 1930s. (They were given their Russian name later.) In World Wars I and II what the flamethrowers really threw was burning gasoline. Napalm, based on gasoline, became infamous during the Vietnam War. In Afghanistan, the United States killed and demoralized Taliban troops with "fuel-air" explosives. These were dreadful for the same reason that the 9/11 attacks succeeded: the huge energy density of gasoline. Seven tons of gasoline, mixed with air and detonated from a parachute, releases the energy of over 100 tons of TNT. So don't drop TNT bombs; you are wasting airplane lift capacity. Carry and drop gasoline. It gives fifteen times the bang per ton.

The 9/11 terrorists did not use high power to destroy the World Trade Center. They used the high energy of jet fuel. The energy released created a high temperature—that is, rapidly moving (vibrating) molecules in the steel of the structure. When molecules shake back and forth, they push nearby molecules away. That's why hot things expand. But that increased separation also weakens the force of attraction between the steel molecules. As a result, hot steel is weaker than cold steel. The weakening of the steel structure eventually led to the collapse of the building.

The 9/11 terrorists took terrible advantage of these facts. When Mohamed Atta boarded American Airlines Flight 11 in Boston, the only illegal things he carried were his intentions—no guns, no explosives, no long knives. Despite all the well-documented weaknesses of our airline inspections, the risks of being caught with a weapon were too great for Atta and his fellow terrorists to take that chance. And they didn't have to.

The genius of the operation was its low risk. No explosives were needed. No illegal weapons had to be smuggled on board. Virtually no organizational infrastructure was required. The danger of the plan being discovered was small because the only terrorists who had to know the details of the mission were the pilots. Atta's plan depended on an airline policy—a policy in force on that day, but never again—that encouraged pilots to cooperate with hijackers. Don't argue, don't threaten; just do what they ask. That approach had saved lives (and airplanes) in the past.

Atta and his co-terrorists took early flights, minimizing the risk that they would be behind schedule and making it easier to attack New York and Washington simultaneously. Even more importantly, they took transcontinental flights. That was to make sure the planes were fully loaded with fuel.

Atta knew that September 11 would be the last day on which an airplane could be hijacked with ease. After September 11, sky marshals are hardly necessary, because no pilot ever again will willingly hand over the controls to a terrorist. Even if a hijacker kills the pilots, the courage and fury of the passengers and crew will be unleashed. That's what happened just an hour and a quarter after the World Trade Center attack, when the passengers of United Flight 93 rushed the cockpit.

Inspecting Passengers

The September 11 attacks took advantage of the terrorists' basic knowledge of airport security. Theirs was not a deep or sophisticated knowledge; it was familiar to almost any technically competent person who had read about prior hijackings. Recall what the security procedures were like prior to 9/11. When you boarded an airplane, your carry-on luggage was passed through an X-ray machine. Such a device can spot hidden objects by analyzing their shapes, but the resolution of the X-ray images is not good enough to reveal an object that is cleverly camouflaged. It would not be hard to hide a knife by putting it in a sheath made of the same material but having an innocuous shape. The terrorists probably did not want to use such camouflage, for the small and legal weapons that they carried with them were sufficient. Get caught with a big knife, and security would be alerted that a hijacking was intended. Because several airplanes had been targeted, it was important that nothing suspicious be done until the plane was actually hijacked.

Metal detectors are designed to detect metal knives and guns. They work because metals conduct electricity and most other materials do not. The passenger is asked to walk through a portal that (from the physics point of view) is just a big coil of wire. Electric current is passed through the wire, making it a big electromagnet. This electromagnet induces current to flow in any metal that passes through it, making that metal into an instantaneous magnet itself. Then the coil can sense the presence of that magnet. If you are carrying a permanent magnet, even if made from ceramic (so that it is not a good conductor), the "metal" detector will pick it up too. That's why many books contain hidden magnetic material—so that the detectors at bookstores can detect a stolen book when the thief tries to leave the store.

Metal detectors don't detect knives or guns; they detect conductors and magnets. Because humans are somewhat conductive (mostly because of the dissolved salt in the blood), the metal detector can't be made too sensitive, and that constraint creates a weapon-smuggling loophole. Deadly knives can be made of ceramic (typically zirconia, also used for counterfeit diamonds), and they would pass through undetected. Today, even guns can be made of ceramic, although most such guns have a detectable metal barrel that would trigger the metal detector or show up in the X-ray machine.

The 9/11 terrorists, however, had no need to smuggle high-tech weapons, or even to hide low-tech ones. They took advantage of the fact that security regulations at that time allowed a passenger to carry a knife on board as long as the blade was less than 4 inches in length. This was an arbitrary rule; if you allow a 4-inch blade, why not a 10-inch blade? The regulation had been a concession to people, such as Boy Scouts, who carry pocketknives as a matter of course. In fact, prior to 9/11, most experimental physicists (like me) carried a knife at all times, typically a Swiss Army knife with several blades, two screwdrivers, and a corkscrew. It was a habit derived from the numerous uses that such a knife has (especially the corkscrew).

The terrorists chose instead to carry box-cutting knives, which have short blades but are typically much sharper than pocketknives, almost as sharp as razor blades. They are also better weapons because they don't fold like pocketknives, so they won't accidentally collapse when used to stab. (Locking blades, such as are used in switchblades and "gravity" knives, had been made illegal for just that reason.) The blade of a box cutter can be retracted into the handle, making the knife look innocuous until it is used. Box cutters don't even look like weapons, but just like a tool that an art student might carry. They were a smart choice. And prior to 9/11, they were totally legal on airplanes.

If airport security personnel think something is suspicious (for example, the X-ray operator sees something unfamiliar in your carry-on luggage), they might search you—but more likely they will just send you to be tested by a *sniffer* device. The operator will take a cotton-tipped swab and rub it on your luggage or on your clothing. It is placed in a box that has a system for recognizing the most common explosives. If you have built a bomb, then the vapors from the explosives will probably be found. Sniffers don't have to detect the bomb itself, and they may not be able to if it is well wrapped. But it is very hard to get rid of all the bomb smell from your clothing, hair, fingernails, and other things you carry with you. Most people don't smell like bombs. Terrorists who have been working around explosives do.

Knowing all this, the terrorists planned an attack that would not require using explosives or a detectable weapon. They probably just sent their box cutters through the X-ray. Or maybe they handed the box cutters to the inspectors, who then would have handed them back after the terrorists had passed through the metal detector. Airport inspectors did not fail on 9/11; it was the security policy that failed. It was our anticipation of the nature of the likely terrorist attack that was deficient. Americans did not expect any-one sophisticated enough to hijack an airplane to be willing to commit suicide.

Taking Control

After the flights took off, the terrorists took control of the airplanes. This was so easy that they did not need weapons to do it. As mentioned earlier, in 2001 it was airline policy to cooperate. Why didn't the airlines anticipate a suicide attack? Certainly there were warnings and fictional accounts describing how that might happen, but it is virtually impossible to prepare for everything described in fiction. Warnings are easy to give but tough to act on. The security people were well prepared for the kinds of attacks they had experienced in the past—the kind they considered more likely. In 2001, most people in the United States were quite pleased with the way prior hijackings had been handled. On several occasions, hijackers had demanded that a plane be flown to Cuba, and that had been done. In each case, when the plane landed, Fidel Castro promptly arrested the hijackers and threw them into a Cuban jail. Castro was not a friend of the United States, but being the destination of all hijackers was not good public relations. Castro promptly returned the airplanes to the United States.

The 9/11 terrorists could have taken control of the airplane by walking through the cockpit door (which, prior to that day, was usually left open). If they said that they had weapons, the pilot would take them at their word. There was no reason to risk trying to smuggle guns on board. Guns would have been superfluous.

Flying and Navigating the Airplane

Most airplane pilot training consists of learning how to land the plane, how to take off, how to make sure nothing goes wrong, and what to do if something does. Flying a plane level is relatively simple. You already know this if you've watched any of the various airplane disaster movies. I've flown in several small planes with my physics mentor and amateur pilot Luis Alvarez. Right after takeoff, he lets me take control. I've flown planes for hundreds of miles with no more safeguard than having a pilot sitting next to me—a pilot who for several hours had no need to make a comment to me. A commercial airplane may be more difficult to fly, but the hijackers had been to flight school. Keeping an airplane in relatively level flight, even aiming it at an airport runway, is not difficult. Landing is the hard part, at least if you want to stay alive.

Navigation is likewise relatively easy, especially if you are aiming for tall structures such as the towers of the World Trade Center. Even to fly a small airplane, you must learn how to use the navigation equipment, and several of the terrorists had training. They also could have used a simple GPS system in the cockpit. In 2001, such systems could have been purchased for less than $200; now they are cheaper. GPS can tell you where you are, how fast you are going, what direction you are traveling, and the distance to your goal (for example, the World Trade Center, keyed in ahead of time). I've used a GPS system successfully from an airplane passenger seat, holding it near the window. Such capability depends on having at least three satellites in view. The GPS measures the distances to those three satellites, and then computes the unique location on the Earth that has those distances.

Of course, the terrorists could also depend on the most ancient of navigation methods: landmarks on the ground. The hijackers of Flight 11 may have found New York City simply by flying down the Hudson River.

In contrast, if you have flown over Washington DC, you may have noticed how difficult it is to pick out landmarks. Even the Washington monument is difficult to locate. From the air, it seems very small. Even the White House is tiny and very tricky to spot. But the Pentagon is different. It is huge and unique and an easy target. The problem with hitting the Pentagon is its low profile. To hit the

World Trade Center, all you have to do is aim the airplane, at almost any altitude; in contrast, hitting the Pentagon requires getting the altitude right. That's hard to do, and it appears that the 757 airplane of American Airlines Flight 77 hit the ground first and skidded into the side of the building.

The airplane that hit the South Tower of the World Trade Center was steeply banked as it hit; this angle is evident from the video taken from Battery Park. Some people have stated that the banking shows that the pilots were highly trained. That's ridiculous. It shows that the pilot had aimed the airplane badly and was trying a last-minute desperate maneuver to hit the building. Unfortunately, the maneuver succeeded.

Impact

When the airplane smashed into the World Trade Center tower, it destroyed a significant number of columns along the perimeter that supported the upper part of the building. The windows and outer wall were ripped apart, the airplane broke into pieces, and much of the 60 tons of fuel, stored mostly in the wings, spewed out. But enough of the perimeter columns of the tower remained to continue supporting the upper floors. They survived. The building was well designed.

Many things could have ignited the fuel, including energy from the impact (sudden compression ignites the fuel in a diesel engine), sparks, or just heat from the engine. Crashed automobiles frequently burn, particularly if their gas tanks are ripped open. But they don't explode, despite what current action movies would have you believe.

There was no time for the fuel to mix well with air, so the fireball was not really an explosion, at least not the kind that does enormous extra damage. In a true explosion the high-pressure gas expands with such force that it can shatter concrete. In the World Trade Center fireball, the expanding gases were subsonic (slower than sound velocity), and mostly they just passed around the support columns, as subsonic gases will do. They blew out walls that were not load-bearing but left most of the structure intact. In the technical jargon of science, we might call it a *deflagration* rather than an explosion. That's why no great booming bang was heard on the street, but mostly a loud roar. Some people in the building did report small explosions; those might have been from fuel that was ignited in rooms or elevator shafts or other confined spaces.

The steel columns of the World Trade Center were covered with insulation and were designed to maintain their strength for 2 to 3 hours of ordinary fire. However, the material that burned was not just office furniture and paper documents; it was airplane fuel. The rate of burning was limited by the availability of oxygen, not of fuel. Low air means a longer burn, and that gave the heat more time to penetrate the insulation.

At high temperatures, steel melts. At much lower temperatures, it softens. At lower temperatures still, it just weakens. All of these changes can be attributed to the shaking of the molecules, driving the iron atoms farther apart and loosening the bonds that give steel strength. The jet fuel created a holocaust far hotter than had been anticipated by the building designers. They never imagined having tons of jet fuel adding to the potential fire problem and burning for over an hour near the top of their skyscrapers.

When the temperature of the columns reached 1500°F, they weakened enough to become vulnerable to buckling. Buckling is not a gentle collapse. Take a paper drinking straw, and push the two ends toward each other. The straw resists with remarkable strength, considering that it's just paper—and then very suddenly collapses. That's because the strength is very high under pure compression. But when the straw (or column) bends a little bit, it buckles easily. Paper has very little strength against bending.

As soon as one column of the tower buckled, it no longer supported its share of weight above it. All the weight now rested on the remaining columns, and they had been weakened too. Probably a second column—one close to the first—buckled at that moment.

Figure 1.1. Collapse of the South Tower of the World Trade Center, September 11, 2001, after being struck by an airplane hijacked by al-Qaeda terrorists.

Then another; the result was like an avalanche. As more columns buckled, the rest could not possibly support all the weight. The entire floor collapsed in less than a second. The massive part of the building above then fell like a sledgehammer, slamming into the floor below. Take another look at the collapse, shown in Figure 1.1. The tilting of the upper part of the building reflects the fact that the columns on one side buckled before those on the other.

Such a collapse multiplies force in the same way that an ordinary hammer does. Let's review that for a moment. As you swing a hammer, you are applying a force to it over a distance of perhaps a foot. You do that to accelerate the hammer. When the hammer finally strikes the nail, the nail absorbs all that energy over a much smaller distance, maybe a half inch. The force on the nail, then, is the ratio of these two values: 12 inches versus ½ inch, a factor of 24 times greater than the force you applied to the hammer during the swing.[2] That's what I mean by a force multiplier. In hard wood, the nail might be driven only a tenth of an inch. For that case the force multiplier is 120 (12 inches versus $\frac{1}{10}$ inch). You actually put more force on a nail in hard wood than you do in soft wood—but you do it for a shorter time.

When the upper floors of the World Trade Center slammed into the lower floors, force was multiplied in a similar way. The weight that the upper part of the building put on the floor below it was multiplied by a substantial factor, perhaps 24 as in the hammer example. The lower floor could not possibly support such a force, so its columns buckled too, virtually instantly. Then this floor joined the upper part of the building in slamming into the next lower floor, and so on. The building was crushed, one floor at a time, but almost at the speed of free fall. Each floor was demolished as if it were being smashed by a huge pile driver, and the mass of the pile driver increased as more floors were added. Watch the videos of the collapse and you'll see this happening.

It is unlikely that the terrorists anticipated that this would happen. They might have thought the impact of the plane would topple the building, or maybe knock the top off. More likely, they just thought a high skyscraper fire would be horrific. Certainly the firemen did not anticipate this mode of collapse (and firemen study burning skyscrapers), or they would not have set up their headquarters in the bottom floor of the burning building. In the 1974 movie *The Towering Inferno*, there is no sense that the entire building might fall. The complete collapse of a tall building was essentially unanticipated. We might still be debating the probability that such a thing could happen again, if not for the matched collapse of the second tower.

One of the engineers who designed the building watched the fire on TV and suddenly foresaw this new failure mode. But in the chaos of the morning, he was unable to reach the fire crews to tell them to evacuate the lower part of the building.

When the building collapsed, it brought down with it whatever jet fuel had not yet been consumed. The continued burning of this fuel caused the collapse (again, because of weakening columns) of the nearby Building 7.

It was not an explosion that caused the World Trade Center buildings to collapse. It was not the airplane impact. It was the fire.

2

TERRORIST NUKES

MANY people fear that the next al-Qaeda attack will be nuclear—
either an atomic weapon or a *dirty bomb* (one that spreads radioactivity). An atomic bomb can be amazingly tiny, small enough to be
carried by one person. It could be smuggled into the United States,
perhaps in a small boat or airplane, or maybe in a shipping container.
Stories have persisted that the Soviet Union developed and manufactured a small "suitcase" bomb that could be smuggled into the
United States for a surprise attack. In the early 1960s, the United
States did build a small nuclear weapon called the *Davy Crockett*. It
weighed about 50 pounds and yet could explode with the energy of
several hundred pounds of TNT. It could be carried by a single soldier and launched from a recoilless rifle. Figure 2.1 shows this
weapon, along with the results of a test done in 1962 (before we
ended atmospheric testing of nukes).

The relatively big punch of the small Davy Crockett weapon is
made possible by the fact that a nuclear chain reaction releases typically 20 million times as much energy as an equal weight of TNT.
The largest man-made nuclear explosion was a test carried out by
the Soviet Union on October 30, 1961. It released the energy of 50
megatons of TNT, 3000 times as much as the Hiroshima bomb. Such

A

Figure 2.1. The smallest nuke: the Davy Crockett nuclear weapon. (A) US officials examine the device. (B) The Davy Crockett is exploded in a test in July 1962.

B

a device could destroy all of New York City. For many of us, fears of such catastrophes provide our reference for nuclear weapons.

Not all nuclear weapons, however, are that big. The Davy Crockett explosion in Figure 2.1B looks horrific, but notice that nothing in the image defines the scale. In fact, a quarter-kiloton explosion from the Davy Crockett is enough to devastate Yankee Stadium, but not much more. The Davy Crockett was deployed on the border between East and West Germany, and in the demilitarized zone between North and South Korea. It's purpose was to repel an invading enemy without turning the border into a no-man's-land.

The Davy Crockett was designed by Ted Taylor, one of the true geniuses of nuclear weapons design. To create this weapon, however, required not just genius but a deep understanding of nuclear weapons physics, exquisite computer analysis, and the results of a large testing program. Nobody knowledgeable in the design of nukes believes that such a weapon could be produced by a terrorist group.

What could terrorists do? Every few years, we hear of a high school student who has "designed" a nuclear weapon. The design is shown to an expert at one of our nuclear laboratories, who is asked if it will work. The answer is, invariably, "I can't say that it wouldn't." The newspaper typically reports that answer as yes.

To put that in perspective, imagine a high school student's design of a supersonic airplane. The sketch has arrows labeling jet engines, swept-back wings, and a cockpit. A newspaper takes this picture to an aeronautical engineer and asks, "Will it work?" The answer is, invariably, "I can't say that it wouldn't."

A sketch is not a design.

According to some experts, a sophisticated group of terrorists *could* make a weapon in the kiloton range, provided that they could obtain the special nuclear materials: 10 pounds or more of bomb-grade uranium or plutonium. Getting that is not easy, as I'll explain in the nuclear weapons chapter.

What is meant by a *sophisticated* group of terrorists? Such a group would include a (probably disgruntled) physicist who had worked in a nuclear weapons program. It would need PhD-level engineers with extensive experience in the behavior of materials under explosive conditions. They would have to be joined by expert machinists and technicians. We can't rule out the terrorists putting together such a team, but it is important for a future president to recognize that we are not talking about people in the Mohamed Atta category.

I know many of the secrets of nuclear weapons (none of which I reveal in this book), and I have been shown the design in detail by the people who designed them. My own evaluation is that a nuclear weapon designed by anything less than a top-level team is likely to fizzle. On October 9, 2006, North Korea tested its first atomic bomb. Though a poor country, North Korea assembled the enormous resources to do this. The yield of this bomb was less than 1 kiloton, and essentially all the experts I know agree that it was probably designed to release 20 kilotons or more. It was a fizzle.

Nevertheless, let's examine what would happen if that North Korean weapon were smuggled into a major US city, San Francisco, and exploded. Figure 2.2 shows the sort of devastation that would occur. The inner circle identifies the radius of destruction. The

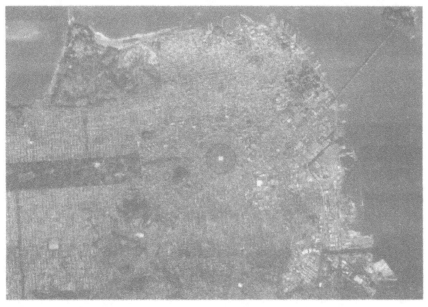

Figure 2.2. Projected effects of a 1-kiloton blast in San Francisco. The inner circle is the region destroyed by the blast; the outer circle is the region with deaths from flying debris.

outer circle shows the area that would have included substantial numbers of deaths from flying debris. The figure was calculated using the Federation of American Scientists' online Special Weapons Primer (www.fas.org/nuke/intro/nuke/effects.htm).

Are you surprised at how small the circle is? If so, it's because you are used to thinking about the large doomsday bombs, the kind carried in our nuclear warheads and bombs. A 1-kiloton blast radius is about 450 feet. If the bomb were detonated in the middle of New York City's Central Park, the blast would not destroy buildings outside of the park. Nuclear radiation has a longer reach, though, and might reach the surrounding buildings, although it would not penetrate beyond the first row. Most deaths might, in fact, be caused by shattered glass on the periphery of the park.

Do I seem to be trying to minimize the dangers of a terrorist nuclear weapon? I'm not. A blast of 450 feet is huge. It would do damage comparable to that inflicted on the World Trade Center. Recall that the energy released from gasoline in that event was about 1.8 kilotons of TNT equivalent, substantially larger than the North Korean nuclear blast. However, a small terrorist nuke would not be much worse than that. A president must recognize that other types of terrorist attacks—some far more accessible to terrorists than nuclear weapons are—can produce casualties of equal or greater horror.

The death and destruction depend on time and place. A 1-kiloton blast in the center of the Rose Bowl on New Year's Day could kill 100,000 people. But such a blast in the middle of New York Harbor would destroy nearby boats, and virtually nothing more.

Big Nukes

Big nukes are a different matter. Each warhead carried on our nuclear missiles is no larger than a man, yet it delivers 100 kilotons of TNT equivalent. The blast range for such a bomb is over half a mile, and the thermal radiation extends 2 miles. The M83 bombs carried by B-52s had megaton warheads, with a blast radius of 2 miles and thermal effects extending to 5 miles. Such a bomb could destroy the entire lower half of Manhattan.

Terrorists are unlikely to be able to manufacture such a weapon. As I'll describe in the nuclear weapons chapter, the only way to achieve such huge explosions is to use two-stage bombs, with a nuclear primary that ignites a hydrogen secondary. Every expert I know agrees that such a design is beyond the capability of any terrorist group. The huge scientific and engineering program required to do it may even be beyond the capability of many countries.

The real worry about nuclear weapons comes from the danger that they might be stolen and then sold to a terrorist group. Here's the scary scenario: A warehouse worker whose job it was to store nuclear weapons in the old Soviet Union squirreled away a few of the weapons when that country collapsed. He has a few of these that he will sell to the highest bidder. Al-Qaeda pays him a visit.

Fortunately, it is not clear whether al-Qaeda would be able to detonate such a weapon successfully, even if they had it. Both the Soviet Union and the United States have elaborate systems to prevent unauthorized people from setting off their weapons. According to Luis Alvarez, one of the physicists who participated in the Manhattan Project (the US nuclear weapons program of World War II), even the Soviet weapons are designed to destroy themselves if they are tampered with—not by nuclear detonation, but by a conventional explosion that destroys the bomb design.

A stolen weapon still remains the greatest threat. The main defense against this threat is to make sure that we, the United States, have lots of undercover salesmen out there pretending to sell real weapons to the terrorists, and undercover buyers bidding to purchase weapons from those who might have them. Russia undoubtedly has similar operatives. In this difficult and dangerous market, we need to make it extremely hard for real buyers and sellers to find each other.

Many people think a dirty bomb is an easier alternative for terrorists. It may be easier, but it is just as likely to fail.

The Terrorist Dirty Bomb[3]

Radiological weapons—the fancier name for dirty bombs—do not require nuclear explosions like fission or fusion weapons do, but instead use ordinary explosives to spread preexisting radioactive material. Saddam Hussein reportedly tested such a weapon in 1987 but abandoned the effort when he saw how poorly it worked. In 1995, Chechen rebels buried dynamite and a small amount of the radioactive isotope cesium-137 in Moscow's Ismailovsky Park. They then told a TV station where to dig it up. Perhaps they recognized the truth: that the bomb's news value could be greater if it were discovered before it went off. The psychological impact of such weapons can be greater than the limited harm they are likely to cause.

José Padilla, a onetime street thug who had received extensive al-Qaeda training, had planned to explode a dirty bomb in the United States. According to a US Justice Department deposition in 2004, al-Qaeda had doubted that Padilla's proposal to build a dirty bomb was practical. They directed him instead to blow up two apartment buildings using natural gas. They apparently felt that such an action would have a greater chance of spreading death and destruction than would a radiological weapon. According to physics considerations, al-Qaeda was right. Perhaps that should scare you. Al-Qaeda appears to understand the limitations of these devices better than do many government leaders, newspapers, and even many scientists.

I don't mean to suggest that radioactive materials are harmless. Indeed, consider the story of scavengers in Goiânia, Brazil, who found and dismantled an abandoned radiotherapy machine in 1987. The machine contained 1400 curies of cesium-137. (A curie is the radioactivity of 1 gram of radium, the weight of two chocolate chips.) Two men, one woman, and one child died from acute radiation poisoning; 250 additional people were contaminated. Several of the 41 houses evacuated could not be cleaned adequately and were demolished.

Imagine if such radiation weren't confined to a few houses, but were spread over the city by an explosion. Wouldn't fatalities be higher? The surprising answer is no. If the radioactivity were dispersed in that way, a larger area would have to be evacuated, yet in all probability no specific deaths would be attributable to the event.

To understand the details, let's walk through the design of a dirty bomb similar to what we think Padilla originally wanted to build. I'll assume the same amount of radioactive material as in the Goiânia incident: 1400 curies of cesium-137. Radiation damage is measured in a unit called the *rem* (which is both the singular and the plural form), and if you stand 1 yard from such a cesium source, you'll absorb 450 rem in less than an hour. This dose of radiation is 50% higher than the *LD50* (which stands for *lethal dose 50%*) for cesium-137, meaning that, untreated, you'll have greater than a 50% chance of dying in the next few months from that exposure.

To try to enhance the damage, let's use explosives to spread our 1400 curies over a larger area—say, a neighborhood 1 mile square. The result will be a radioactivity of 0.5 millicurie per square yard, and a careful calculation shows that if you are in this area, then after an hour of exposure your exposure will be 0.005 rem, 5 *millirem*. That's a tiny amount, far below the threshold for radiation illness (100 rem), so you won't get sick at all. If you stay in the area, even after a month your dose will be only 4 rem, still way below the threshold for radiation sickness. There will be no dead bodies whatsoever, unless someone is killed by the explosion itself. I suspect that's why al-Qaeda instructed José Padilla to abandon the dirty-bomb concept and try to plan a natural-gas explosion instead. True, low levels of radioactivity can induce cancer, but that takes years. I suspect that al-Qaeda doesn't just want to brag about the number of premature cancers that will be induced by their attack. They need their followers to see photos of bodies.

Now let's look at the cancer risk. For moderate doses, results from historical exposures suggest that the increased risk of cancer is about 0.04% per rem. Multiply 4 rem by 0.04%, and you get a danger of induced cancer of about 0.16%. If we approximate the cancer rate in the United States from "natural causes" as exactly 20%, then living in the dirty-bomb zone day and night for a year would increase the likelihood of dying from cancer to 20.16%. That's bad, but is it bad enough for you to evacuate your home? (I assume that after one year, the radioactivity could be cleaned up.)

Of course, if the radioactivity is more concentrated, then the danger is larger—but to fewer people. The radioactivity is most intense before the bomb is exploded, and that puts the terrorists at truly great risk. They have to assemble the bomb and deliver it before they are killed by it. Energetic gamma rays are not easy to shield, so the bomb would be assembled probably behind tons of lead. If it were delivered without the lead, it would have to be delivered quickly, before the terrorists dropped from radiation illness. That is not easy to do.

In the 9/11 attacks, the terrorists took advantage of US policy and prejudices. They knew they didn't need guns to take control because pilots had been instructed to cooperate with hijackers; that policy had been based on the fact that nobody had expected hijackers to turn planes into weapons. Similarly, a terrorist today might use a radiological weapon, not because of its actual damage but in anticipation of the out-of-scale panic and ensuing economic disruption that the weapon could trigger. If a radiological attack is mounted, it may be your responsibility as president to convince the public that they must not overreact. That is why you need to understand the physics. The public will not be satisfied if you have your science advisor make all the reassurance speeches.

Could other radiological attacks be more potent than our hypothesized cesium-137 example? Electric generators powered by the decay of radioisotopes, found in abandoned lighthouses in Russia, held 400,000 curies of strontium-90. But strontium-90 emits virtually no gamma rays; it is harmful primarily when inhaled or ingested. A cloud of strontium-90 released in aerosol form can kill—but it settles quickly and does not stay in the air long. Just don't eat the food it settles on, and don't eat the animals that eat the grass it settles on. And don't drink their milk.

For the same reason, even a radiological bomb made with plutonium is unlikely to be dangerous. Anthrax would be deadlier, and much easier to obtain and transport. Nuclear waste storage facilities and nuclear reactors contain vastly more radioactivity, and the danger from them would be substantial if their radioactivity could be transported, released, and spread. We'll talk more about nuclear waste in the chapter devoted to that subject. But all dirty bombs have the same problem: intense radioactivity from the unexploded bomb that can kill the terrorists, and diluted radioactivity after it is exploded that drops below the threshold for radiation illness, unless the area attacked is very small.

If dirty bombs threaten so little harm, why are they lumped in with other weapons of mass destruction? The reason is that it's the law, as written in the 1997 National Defense Authorization Act (Public Law 104-201) and other places, including California Penal Code Section 11417. Defining them like this was a mistake that could lead to misallocation of resources and a general overreaction if such weapons were used.

The biggest danger from a radiological weapon is the misplaced panic and overreaction that it would cause. A dirty bomb is not really a weapon of mass destruction, but it is potentially a weapon of mass disruption.

Rogue Nukes

In many ways, the real threat is not nukes made by terrorists but rather nukes made by *rogue nations*. This term is vague, and many people object to it (is Israel a rogue nation? is India?), but it is a useful concept and certainly a term that a future president will hear. The term is usually applied to relatively undeveloped countries that violate a perceived commitment to the Nuclear Non-Proliferation Treaty and covertly develop nuclear weapons. There are several prominent examples: Iraq prior to the first US invasion; North Korea in recent years; Iran at present. A rogue nation might deny that it is making a weapon, or claim that its technological development has only to do with peaceful applications (so the nonproliferation treaty doesn't apply), and finally (after it has tested the weapon) argue that it had a right to do so because its national security was threatened.

The physics that you need to know on this subject has largely to do with the signs that your intelligence agency might pick up. The dominant worry is that the rogue nation will develop "small" nuclear weapons, such as the ones that the United States used to destroy Hiroshima and Nagasaki in 1945. There is no concern that they will develop a thermonuclear weapon until they have gone through the first stage of building a uranium or plutonium bomb. They might use such weapons in war, though, or provide them to others—for example, terrorists. These two kinds of fission weapons—uranium and plutonium bombs—require quite different technologies, so let's summarize the key things to watch for in these two cases.

Uranium Bombs

The design of a uranium bomb is simple; the problem is obtaining enriched uranium-235. The modern technology for doing this is the gas centrifuge, which we'll discuss in the nuclear weapons chapter. Currently Iran admits to developing a gas centrifuge system and claims that the only purpose is the partial enrichment needed for a uranium-based nuclear reactor. One of the key materials used in modern centrifuges is called *maraging steel*. The importation of steel tubes by Iraq just before the second US invasion was an indication that they might be trying the gas centrifuge method. Back in 1990, Iraq had taken a different approach toward the enrichment of uranium: building devices called *calutrons* (described in Chapter 10, Nuclear Weapons), the same devices that the United States had devised to enrich uranium in World War II.

To detect work on a uranium bomb, be on the lookout for centrifuge components: special bearings needed to suspend them at high speeds, or maraging steel to keep the centrifuge from spinning apart. But don't forget to look for cruder but easier separation technologies, such as calutrons.

Plutonium Bombs

Plutonium is relatively easy to get. Many countries now operate nuclear reactors, and in these devices the abundant kind of uranium called U-238 is gradually turned into plutonium. The Nuclear Non-Proliferation Treaty specifies that the abiding country may not extract this plutonium from the reactor waste. To do so is called *reprocessing* the fuel. To prevent covert reprocessing, nuclear power plants are inspected by the International Atomic Energy Agency. When North Korea refused such inspections, the fear was that they had begun reprocessing. Indeed they had. On October 15, 2002, they announced that fact to US Assistant Secretary of State Jim Kelly. Michael May, a US nuclear weapons expert, visited Korea to confirm that they had reprocessed plutonium. He asked to be given the substance to hold in his hand. He knew there was no danger from the emitted alpha particles, because they would not penetrate the dead layers of his outer skin. He evaluated the heft and warmth of the sample from the radioactive decay. These were good indicators because the North Koreans clearly had not expected him to ask to hold the material unshielded. May's observations led him to conclude that the plutonium was real.

For a rogue nation with nuclear reactors, getting the plutonium is the easy part. As I'll describe in the nuclear weapons chapter, a plutonium bomb requires a carefully tuned implosion, and that is very difficult to achieve. To do so requires not only exquisitely shaped explosive charges and simultaneous detonation, but also an extensive testing program. And even if those succeed, there is no guarantee that the bomb will work when used. On October 9, 2006, North Korea announced that it had successfully tested a nuclear device.

When experts analyzed the seismic signals from the blast, however, they concluded that the yield was so low (less than 1 kiloton) that the explosion was undoubtedly a fizzle, probably because of incomplete implosion of a plutonium core.

To detect work on a plutonium bomb, look for diverted material from a nuclear reactor, a chemical reprocessing plant designed to handle highly radioactive material, and an implosion testing program.

3

The Next Terrorist Attack

A NUCLEAR terrorist attack is possible, but it may not be the biggest threat. Al-Qaeda has shown persistence and focus. Their successful attack on the World Trade Center was their second attempt at destroying that building. It succeeded, in my estimation, beyond their greatest hopes. It is hard to think like terrorists, but let's try. Are they more likely to try something new and exceedingly difficult (nukes) or to use an approach similar to one that has already proved enormously successful?[4] I suspect the latter. There is a very good chance that the next terrorist attack will not use Western high tech, but will leverage our own sophistication against us. Let's consider what al-Qaeda can do with a low-tech attack.

Future Airplane Gasoline Attacks

Gasoline could easily remain the low-tech weapon of choice for terrorists. It is unlikely that they will try hijacking another American commercial jet, largely because such an attack is extremely likely to fail. That was demonstrated by the rebellion of the passengers of United Flight 93, the fourth of the 9/11 planes. Passenger fury was also illustrated by the fate of the shoe bomber Richard Reid, who was prevented from igniting his explosives by watchful and angry passengers.

A gasoline attack from the skies is still possible, however. Consider the history of the 9/11 terrorists. Before his attack on the World Trade Center, Mohamed Atta had made repeated trips to rural airports, trying to learn everything he could about crop dusters. Why crop dusters? Many people assume he was interested in spreading chemical or biological terror over a major city. But there is no evidence that al-Qaeda had stores of such materials. Searches in Afghanistan turned up no major chemical or biological facilities capable of delivering large amounts of biological poisons; nor is there evidence that the group somehow had access to large stockpiles in the United States.

Other facts reinforce the al-Qaeda interest in crop dusters. A computer disc with information about crop dusting was found in the possession of Zacarias Moussaoui, who was convicted of conspiring to commit aircraft piracy and kill Americans in the 9/11 attack. At the US trial of the four men accused in the 1998 US embassy bombings in Kenya and Tanzania, federal witness Essam al Ridi testified that Osama bin Laden wanted to buy a crop-dusting business. Johnelle Bryant, former manager of a farm agency associated with the US Department of Agriculture, said that Mohamed Atta, the leader of the 9/11 attack, had visited the USDA to try to obtain a loan to buy a crop duster and modify it to carry more gasoline.[5]

A terrorist interest in crop dusters makes sense if you think about the physics. An Air Tractor 502 crop duster airplane is far smaller than a 767, but it is also a flying tanker. It has fertilizer containers that hold roughly 320 gallons of liquid, plus a 130-gallon fuel tank. It flies close to the ground, where it cannot be detected by most radar technologies. Fill 'er up with 450 gallons of gasoline, and you are carrying roughly 2.1 to 2.4 tons of fuel—the energy equivalent of 32 to 36 tons of TNT.

What could a single suicide pilot do with a full crop duster? He could crash into Yankee Stadium during the World Series, or into the Super Bowl, or into the Olympics opening ceremony. The deaths, including trampling, might exceed those at the World Trade Center, with everything broadcast live on international TV. (I virtually held my breath during those events in 2002.) Or the pilot might target a petrochemical plant, or a nuclear waste facility near a large city. (That's where most nuclear wastes are now and will continue to be stored, unless facilities such as the one being built at Yucca Mountain in Nevada are approved.)

Fortunately (for us), the Air Tractor 502 is notoriously difficult to fly, particularly when fully loaded, and piloting this plane is even more difficult if the plane is kept at low (radar-avoiding) altitude. That makes it an unlikely terrorist weapon for the future. Maneuvering a Boeing 767 is probably easier, as long as landing is not required. Moreover, the crop duster community is very close, small, and wary. Even before September 11, they didn't let Atta photograph their planes or even sit in the cockpits. I have discussed this issue with the crop duster community. They needed no prodding from me. They are very sensitive to the terrorist threat. For the foreseeable future, you can be assured that every suspicious visit to a crop duster facility will be reported immediately to the FBI.

Don't take too much comfort, however, in the difficulty of obtaining and flying a crop duster. Other kinds of small planes can be used. Gasoline is a low-risk explosive, and purchasing it doesn't require a special license. So beware. If al-Qaeda agents want to kill and terrorize, then their weapon of choice may well be one that can be bought at the corner station.

Bombs on Airplanes

To crash a gasoline-laden plane into a populated structure requires the ability not just to fly the airplane, but also to take off from a runway and (very likely) to land the plane. That's because most takeoffs are practiced using a procedure called *touch and go*, in which the plane lands and immediately takes off again. Moreover, the terrorist would have to buy the airplane and endure the suspicions of the seller. In many ways, an attack using a small airplane is more complex than hijacking a commercial plane. For that reason, the more realistic threat may be that multiple suicide bombers carry hidden explosives on several different commercial flights and attempt to blow up the planes simultaneously across the United States.

Richard Reid was the al-Qaeda terrorist who tried to light a fuse on his shoe bomb to blow up American Airlines Flight 63 on December 22, 2001. A passenger in the airplane took a photo (Figure 3.1) of fellow passengers ganging up on Reid and thwarting his efforts.

The intelligence experts concluded that Reid himself didn't design (couldn't have designed?) that shoe. Was Reid on an official

Figure 3.1. Passengers of American Airlines Flight 63 gang up on terrorist Richard Reid, preventing him from detonating his shoe bomb.

al-Qaeda terrorist mission? I'm guessing the answer is no. Reid had become frustrated at the lack of communications and orders (the al-Qaeda organization in the United States, at least, has been badly broken), and he decided to go ahead and blow up a plane himself.

That was very, very stupid. Al-Qaeda is not interested in blowing up one plane; they want to blow up a dozen. They knew they could smuggle explosive-laden shoes on board, and (I am guessing) they had a dozen of these shoes all set for simultaneous attack. Reid, in his impatience, blew the secret of the scheme. I bet the other eleven (of the dozen, if that was the number) shoes are still out there. But now shoes are checked by airport security, and as long as this is done, the larger plan will be impossible.

That's why airport security makes you take off your shoes. They are not particularly interested in whether the shoes pass the metal detector; a shoe with explosives doesn't have to have metal. They are looking for the other shoe bombs. The US Transportation Secu-

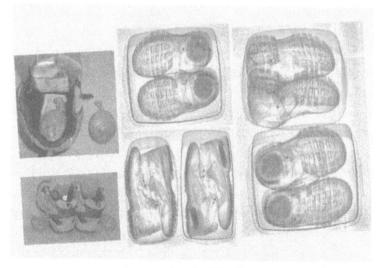

Figure 3.2. X-ray of a shoe bomb. The Transportation Security Administration prepared this photo on the basis of the Richard Reid device. The photo was distributed to security agents to show them what to look for.

rity Administration has circulated a photo (Figure 3.2) showing their security agents what to look for. The little packets are the plastic explosives. It isn't much and, if exploded in the middle of the passenger compartment, probably wouldn't do more than injure a few passengers. To be effective, such a weapon has to be used in a special place within the plane. When Reid tried to light the fuse, he had reportedly wedged the shoe between a window seat and the wall of the airplane, as if he wanted to blow a hole in the structure. That could have led to a disastrous weakening of the airplane structure and a crash.

In August 2006, about two dozen people were arrested in Great Britain and accused of a terrorist plot to blow up airplanes using "liquid explosives." The nature of the explosives has never been officially released, and it is not clear why liquids would have any advantage over solids. Some said that there were two liquids that would not become explosive until mixed. This approach would be of value only if the liquids were such that they would not be detected by the sniffers at airports.

There is no good way to detect carefully prepared explosives. Neutron activation, which detects the nitrogen in explosives, has received the most attention. But this technique generates too many false alarms—typically several per full flight—from leather and other materials that also contain lots of nitrogen. What do you do with luggage that sets off a bomb detector? Open it? Where? Blow it up? There is no good solution, as long as there are abundant false alarms.

Better explosive detectors are under development. Nuclear electric quadrupole resonance—a method that detects the chemical environment of the nitrogen nucleus—offers real hope with few false alarms, but it is not yet ready to install at airports. The best bet today is the ion mobility time-of-flight spectrometer. This is the sniffer that is in wide use at airports. If suspicious, a security agent takes a swab of your luggage—maybe even of you—to analyze in the machine. The device costs less than $50,000 and has a false-alarm rate of less than one in a thousand. But it would miss a carefully wrapped explosive, unless the outside of the package (or the person carrying it) was contaminated.

On a recent trip to France, I was stopped after an X-ray inspector noted something suspicious in my carry-on luggage (probably the bag full of chargers for my video camera, digital still camera, cell phone, iPod, computer, voltage converters, extra batteries, and extension cords). How could he check all these? He didn't. Instead he asked me to take off my shoes, and he put them in a sniffer. Smart, I thought! If I really were a terrorist, there might be residue from explosives on my shoes.

Can't explosives be put in checked luggage? Yes, of course. To address this problem, airports now require luggage on a plane to be matched to the passengers who have actually boarded. This requirement causes all sorts of inconveniences; I was on a flight recently in which they had to unload all the luggage to find the bags of someone who had to change his flight. Some people think that matching passengers and luggage is a waste of effort, since the terrorists can simply use suicide bombers. But that criticism misses the point. *Forcing* al-Qaeda to use suicide bombers gives us a great advantage. It significantly limits the number of people available for terrorist duty. And the ones remaining are hardly *la crème de la crème*, as we once thought.

Let's look at the personalities of the successful suicide bombers. According to the testimony of Johnelle Bryant, the woman who interviewed Mohamed Atta about his crop duster plans, he was hardly a man who could today move unnoticed in the Western world. She said that when she interviewed him, he objected to her because she was "but a female," and then he threatened her life ("What's to prevent me from cutting your throat?" he said). Such behavior would not go unreported today; it amazes me that it did prior to 9/11. Several of the other terrorists were equally inept. Richard Reid couldn't ignite his own shoe. José Padilla, the man given the task of making a radiological weapon, was a former Chicago street thug with a long arrest record. Zacarias Moussaoui couldn't pass a simple written exam in flight school and told his teachers that he wanted to learn to fly big planes but was not interested in taking off or landing. He was reported to the FBI and arrested. Moussaoui even insisted on acting as his own lawyer—as if to confirm his status as a fool. On a suicide mission today, these characters would stick out like rotten thumbs. As long as we force al-Qaeda to use such people, they will be noticed (even if they weren't before 9/11), and that makes a coordinated attack virtually impossible. That's part of the reason why—with the exception of the anthrax mailings, which I'll discuss in the next chapter—there has been no major terrorist attack in the United States since 9/11.

What about the airport security rules against scissors and pocket-knives? Those accomplish little or nothing. The danger is explosives. I wish there were a workable technology to detect them. But until there is, let's force the terrorists to use suicide bombers, and let's spot them at the airport. Don't underestimate the success of the security measures. Who do you know who would have predicted, after 2001, that over six years would pass with no additional terrorist triumph in the skies?

4

BIOLOGICAL TERRORISM

ONE week after the 9/11 attacks, the United States suffered a second round of terrorism. Letters containing enough deadly anthrax spores to kill millions of people were mailed to ABC, NBC, and CBS, as well as to the *New York Post* and the *National Enquirer*. Shortly afterward the United States suffered its first death ever from a biological warfare attack when Robert Stevens, working in the *Enquirer* building, died of the infection. On October 9, additional anthrax letters were mailed to Senators Patrick Leahy and Tom Daschle (see Figure 4.1). Twenty-two people in the United States were diagnosed with symptoms of infection, and five died. The postal service was shut down, and the US government offered a $2.5-million-dollar reward for information leading to the arrest and conviction of the terrorists responsible. As of the writing of this book, we still do not know who was responsible. No subsequent attacks took place, although a spate of hoax letters followed. The World Trade Center attacks are clear and well understood compared to the anthrax mystery.

Why discuss anthrax in a physics book? It turns out that the spread of the spores, and their failure to kill more people, is more related to physics than to biology. This physics is something that a

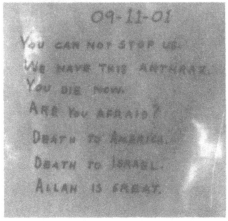

Figure 4.1. The anthrax-laden
letter that Senator Tom Daschle
received on October 9, 2001.

future president needs to know, not only to understand the threat, but also to do a better job at directing the FBI in its attempts to track the terrorists.

Before I begin, however, I feel compelled to add a few words about other biological threats. Many people worry about the spread of an infectious disease. Smallpox has a much greater potential to kill than does anthrax, because it spreads like a chain reaction. If one person infects 10, and each of them infects 10 more, and so forth, the numbers grow exponentially: 1, 10, 100, 1000, 10,000, and so on. In just nine steps, a billion people could be sick. In contrast, anthrax does not spread from person to person but kills only the few people who breathe the initial spores into their lungs. In farms, anthrax spreads its spores only when they are released from decaying bodies, typically from dead cattle out in the field.

Contagious diseases such as smallpox do pose real danger, but I hope that any terrorist thinking about that will recognize that a smallpox attack would not be restricted to the United States. It would invariably spread around the world. In fact, it is very likely that the United States would be one of the nations to suffer the least, because of its excellent health care system. The true victims of the unleashing of smallpox would be the developing world.

The Anthrax Attack

Many people think the fall 2001 anthrax attack was successful— that it killed roughly the number of people intended: five. It is just as reasonable to conclude, however, that it was a fizzle.

Prior to 9/11, it was generally thought that a very small amount of anthrax spores would trigger true horror. A study posted[6] on the Web by the Defence Research Establishment Suffield in Alberta, Canada, on September 1, 2001—well before the anthrax letters were mailed—suggested that envelope-borne anthrax spores could be released in aerosol form very effectively by the simple act of opening the envelope. The report stated that anthrax dispersal from letters was "far more effective than initially suspected"; greater than 99% of the respirable aerosol particles in an envelope were released into the air when test envelopes were opened. The report concluded that lethal doses can spread rapidly throughout a room where an anthrax-laden envelope is opened. Any terrorist checking the Web in early September might have found this report and decided to act on it.

Consider the amounts of anthrax used in the fall 2001 attack. Virtually all the articles and data in the public domain at that time stated that even a few grams of anthrax (the weight of a penny) could, if dispersed efficiently, kill millions of people. Any terrorist who put this information together with the Canadian study might have concluded that the mail was an ideal way to kill hundreds, maybe thousands of civilians. If that is what the terrorist thought, then the attack was not intended to be a demonstration; it was not planned to disrupt the mail, or even the US economy. It was intended to commit mass murder, including among its victims US leaders and media personalities.

If mass anthrax deaths were the terrorists' goal, how did their plans go so wrong? Here's one possibility: that the terrorists were influenced by the misleading technical concept of *lethal dose*. Consider the following paradox: Senator Patrick Leahy, after a briefing on the possible contents of the letter sent to him, announced on *Meet the Press* that it might contain "100,000 lethal doses." Yet only five people died from all the letters. Was Leahy exaggerating? No. In fact, he was being conservative.

How can we reconcile five with 100,000? Here are the numbers. On the basis of primate experiments, the Defense Intelligence Agency estimates that 2500 to 55,000 spores are enough to trigger fatal pulmonary anthrax infections in half of those exposed (this is the LD50 dose). A single spore could trigger the disease, but the probability is low and on average many spores are required. Ninety-four-year-old Ottilie Lundgren, the fifth and last anthrax victim, may have been killed by just a few spores. That would explain the absence of detectable anthrax in her home and belongings.

To reach the most sensitive areas of the lungs, the spores or clumps of spores must be small, with a diameter not much larger than 3 microns, about one-tenth the thickness of a fine human hair. Leahy's letter was reported to contain 2 grams (about the weight of a penny) of finely divided anthrax with 200 billion such particles. If we assume that 10,000 particles is a reasonable average for LD50, then the letter contained 20 million lethal doses. So Leahy's estimate of 100,000 was actually quite low.

In the worst-case scenario (or the best-case scenario—from the terrorists' point of view), the anthrax powder would sprinkle out of the envelope, disperse like dust, be swept up into a building's ventilation system, and be mixed and uniformly diluted in the recirculating air. A human breathes about a cubic yard of air every hour. With 10,000 particles in each cubic yard—enough that everyone exposed would die—200 billion particles from one letter could (in principle) contaminate 20 million cubic yards, almost the volume of the entire New York City subway system. No wonder people were frightened by the anthrax threat!

This worst-case scenario, however, is highly misleading. The primary challenge in the military use of anthrax has always been to find methods of mixing the spores thoroughly with the air and keeping them suspended long enough to be breathed. Most dispersal methods are extremely inefficient. Lethal doses, per se, aren't meaningful.

The terrorists may not have appreciated this subtlety. Let's assume they had only a few grams of anthrax. They correctly estimated that they had several hundred million lethal doses. Even at only 1% efficiency (a conservative estimate, they mistakenly thought), they could kill 2 million Americans. Of course, the lethality might be limited to one building, and maybe part of the surrounding area, so only thousands would die—or only hundreds, if the terrorists were very unlucky. If that was their thought process, then they grossly underestimated how much luck was needed.

If this scenario is correct, the terrorists must have been surprised that their initial anthrax attack was a failure. Only one person died: Robert Stevens, a photo editor at the *Sun*, a tabloid newspaper. The Canadian test, thankfully, may not have been a good predictor of how anthrax spores behave in the real world. In that study the anthrax was folded in a sheet and was ejected when the sheet was pulled and opened. Maybe no such sheet was used by the terrorists; perhaps they just dropped the anthrax into the envelope, where it remained. Or anthrax might have migrated out of the sheets during the extensive handling by the post office and settled to the bottom. Or, finally, maybe the anthrax did disperse, but only through the rooms where the letters were opened; the Canadian tests did not include measures of dispersal through ventilation, and this kind of dispersal may not be very efficient. In the tests, the half-life of anthrax exposure in the chamber was about 5 minutes, suggesting that the spores settle quickly to the ground, where they are virtually harmless (unless people eat the dirt on the floor). Five minutes is long enough for people in the room to be infected, but not for spores to migrate very far.

When only one person died, the terrorists may have panicked. They had failed in their mission, and they didn't know why. They incorrectly guessed that their anthrax had lost its potency, and in desperation they mailed out all of the remainder, much of it in pure undiluted form, on October 9. That fact would explain why there were no additional anthrax attacks after that date.

Anthrax spores were eventually detected not only at the US Senate and the House of Representatives, but also at the White House mail facility, the Supreme Court, the CIA mail facility, the Pentagon, and all over Washington DC. The prevailing belief is that cross-contamination in the mail rooms was the reason that the spores spread to so many sites. But I think it worthwhile to consider the possibility that some of the detections were the result of early, diluted letters. In their first mailings, the terrorists assumed that the anthrax would spread to more locations.

The FBI spent most of its effort looking for an American who wanted to scare the public by killing a small number of people. Its agents failed to find such a person. I regard it as more likely that the anthrax was stolen by someone with access to American samples, maybe someone who had the job of destroying it—but took it home instead. It would not have to be a scientist, but only a technician who ran the autoclave.

My suggested scenario may seem complex, but real scenarios always are. Not all of the details will be correct. No scenario presently explains everything, and to make sense of the complicated situation requires judging the evidence. Who is more credible when the conclusions conflict: a handwriting analyst who says the terrorist was American; or a medical doctor, Dr. Christos Tsonas, who treated the leg of Ahmed Alhaznawi, one of the September 11 hijackers, and says that "the most probable and coherent interpretation of the data available" is that the infection was skin anthrax?

Is it conceivable that the anthrax attack was the "second wave" planned by al-Qaeda? According to the October 27, 2001, *Washing-*

ton *Post*, a senior official said "nobody" then believed the anthrax attack was the second wave. "There is no intelligence on it and it does not fit any [al-Qaeda] pattern." But whether it fits the al-Qaeda pattern depends, in part, on the intended scale of the carnage. It may be a mistake to assume that the attack was supposed to kill only five people and that it worked as planned.

If I am right, the terrorists may now be disillusioned with anthrax attacks. But it would be foolish to relax. Osama bin Laden was building laboratories in Afghanistan that, given time, could have produced not just grams, but many pounds, of spores. Tons of anthrax was grown in Soviet laboratories and buried on Vozrozhdeniye Island in the Aral Sea (just north of Iran and Afghanistan)—possibly along with some smallpox virus. (It was treated with bleach, but tests show that much of it is still viable.) The Soviet anthrax was reported to be resistant to most antibiotics. So, despite the limited casualties of this first biological warfare attack on the United States, the prognosis is bleak. Biological terror is likely to prove more accessible and easier to implement than nuclear terror. The "mad scientist" of future fears is more likely to be a biologist than a physicist. Even though I am a physicist, that thought does not give me much comfort.

TERRORISM: PRESIDENTIAL SUMMARY

TERRORISM is perhaps your trickiest problem. As time goes on, if there are no further successful terrorist attacks in the United States, US citizens will become complacent. They will depend on you to be vigilant but won't want you to spend much money doing so, and they will expect you to be very respectful of human rights. It's a tough job you have chosen.

It won't be easy for the terrorists. The prior attacks required terrorist groups to maintain an elaborate infrastructure within the United States, and citizen awareness has been heightened to the point that such groups are likely to be noticed and reported. Airport security measures, many of which serve no realistic purpose (such as the confiscation of small knives) may in fact help keep citizens alert. For these reasons, it is likely that the next successful attack will not be in the United States.

Where will you put your focus? Many will advise that the big danger comes from the big threats such as nukes smuggled in shipping containers. Some people argue that the United States must have a sophisticated X-ray or nuclear detector examine every such box that enters this country. A program as ambitious as that is probably unrealistic. Even if such detectors could be built (and the national laboratories are busy trying to design them), a few simple tricks of camouflage could probably defeat them. Don't hide your nuke in a box of hay; rather, hide it in a box of tractor parts. Amid that clutter, the weapon may be overlooked.

People worry about nukes because they could represent a true escalation of the terrorist capability. But they are hard to make, hard to steal (especially if the United States has an active clandestine program to try to sell phony nukes), and hard to detonate. Remember, a small nuke in the middle of New York Harbor would probably do little damage. Dirty bombs are even less fearsome. My recommendation is to put less emphasis on such a high-tech threat. The real danger comes from the Ewok-like attack: gasoline-filled airplanes used as missiles, or shoe bombs for a simultaneous attack on our airlines.

To be candid, I must admit that many of my informed colleagues disagree with me. They think the emphasis should be on preventing a nuclear attack with a smuggled weapon. I think they are wrong. But this is not really a physics or a technical judgment; it has more to do with understanding the terrorist psyche. You have to tap your other (nonphysics) skills to make the final judgment. But remember, your advisors on terrorism may not appreciate the difficulty of making a nuke, so they, too, may have an exaggerated fear.

I suspect that another gasoline attack is likely, in part because it worked so well the first time (both World Trade Center buildings fell, not just one), it causes high numbers of immediate deaths (needed by the terrorist groups to prove to their home people that they can do miracles), and it makes dramatic films for TV. The attackers probably will not use a commercial airliner (security measures make it too risky), but they may use a medium-sized private plane, perhaps a DC-3, loaded with fuel, aimed at something like a football stadium on a Sunday afternoon. The best way to prevent such an attack is for the community that deals with such aircraft to be vigilant.

Another likely attack is to blow up many passenger airplanes simultaneously. More shoe bombs like the one that Richard Reid attempted to detonate are probably out there, and they offer a real danger. Such an attack would require the coordination of a large group of people in the United States, and it is best countered by good police work, and by taking reports of suspicious activity seriously. Continued checking of shoes at airport terminals is a good idea, but metal detectors are insufficient. The shoes must be X-rayed. Don't worry about knives and other sharp objects; they are virtually useless for an airplane attack because the passengers won't allow a hijacking. I would deploy lots of sniffers—mass spectrometers that can detect the vapors of explosives. You are not likely to detect the explosive itself, since it will be carefully wrapped, but you are likely to find residuals on the person carrying it, or on that person's suitcase. Keep the cockpit door locked.

The greatest unknown in my mind is the danger of a biological attack. Making anthrax or mutating a bacterium is relatively easy compared to assembling or even operating a nuke. I suspect that anthrax will not be used, because of the evident failure of the 2001 anthrax attack to kill large numbers of people. If another anthrax attack does occur, make certain you don't overreact to reports of huge numbers of "lethal doses." A greater danger would be the release of a virus or bacterium—maybe one from a remote region of the world, one that has been genetically engineered to be more dangerous. Students with a master's degree or less in biology could know all the procedures necessary to make such a bug. A terrorist who spreads a disease runs the risk of killing more people in the developing world than in the United States, but some terrorists may not care. Good data collection methods would be essential to locate the source of the spread and to isolate it. Such procedures may serve a dual purpose if a "natural" pandemic such as bird flu begins to spread.

Most of the unknowns related to future terrorism are not physics questions. They have to do with the terrorist mind, the possible fears and reactions of our own people, probabilities, weighed risks and costs. As president, you need to have a proper sense of the physics aspects: how hard is it to make a nuclear bomb, how bad is a dirty bomb, what are the dangers of high explosives and gasoline, and what is the threat of a biological attack? You will have to allocate resources, and your decision on how to do this will depend on numerous other issues, many of which are not technical. That's why we have a president in charge and not a physics professor.

II

ENERGY

COUNTRIES *go to war over energy. When the Japanese invaded China in 1941, we responded with an embargo on their oil imports. Once we had done that, many experts believed that an attack from Japan was inevitable. Pearl Harbor followed later that year. In the 1970s the oil cartel OPEC embargoed the United States, creating a crisis felt by every US citizen. We are currently at war in Iraq, in part to bring freedom to the Iraqi people, but also because Iraq is the keystone of the Middle East oil region.*

Energy is important because it is linked to national wealth. As an astonishing example, consider the United States and China. We have a gross domestic product 20 times theirs. Our energy use is also about 20 times theirs. Wealth seems to equal energy use. The correlation is truly amazing—and something that China has noticed. No wonder the Chinese are frantically building new power plants, averaging over one new giant (gigawatt) plant per week.

As a future president, you undoubtedly ponder the future of the developing world. The economy of China is small per person but large in total, and growing rapidly. Most caring people in the world are happy to see the growth of the developing countries, with the associated reduction of poverty, poor health, and poor opportunity that has plagued them. As they grow, however, their energy use will undoubtedly grow too. What about their pollution? In particular, we worry about their emissions of carbon dioxide, the notorious greenhouse gas that is linked to global warming and to acidification of the oceans. China probably passed the United States in the total of such emissions in 2007. Moreover, China emits far more greenhouse gases per GDP than does any other country in the world— over 3 tons of CO_2 for every thousand dollars; US emissions are nearly six times lower. After China, the second worst polluter per GDP is India, emitting nearly 2 tons of CO_2 per thousand dollars GDP. It is absolutely critical that these pollution rates drop as the GDPs of these countries grow.

If their economies expand (as we hope they will), then soon the developing countries will dominate in pollution. What can we do? To answer that question, we must first have a realistic picture of energy—what it is, where it comes from, and where it goes.

5

KEY ENERGY SURPRISES

No FIELD of physics can beat energy for the number of things that *folks know that ain't so*. Let me illustrate this in a positive way with some key energy facts that surprise most people and that are important for making good policy decisions:

- Gasoline delivers 15 times the energy of an equal weight of TNT.
- Coal is 20 times cheaper than gasoline, for the same energy.
- A square mile of sunlight at midday could provide a gigawatt of electric power[1]—the same as a large coal, electric, or nuclear power plant.
- A square yard of sunlight delivers about a horsepower when it hits the ground, the same as the average electric power used by a US household.
- Gasoline has 1000 times as much energy as an equal weight of flash-light batteries, and 100 times as much as an equal weight of expensive computer batteries.
- Liquid hydrogen, the key fuel for a future "hydrogen economy," has 4.5 times *less* energy per gallon than gasoline has.
- Energy from nonrechargeable batteries costs about 10,000 times as much as from the wall plug.

Future presidents need to know, understand, and feel comfortable with these facts, just as much as they need to know the difference between Shiite and Sunni, or the history of the Japan–China conflict. We'll begin with the most important issue: the physics reasons why we love gasoline so much.

Why We Love Oil

It's a cliché that Americans have a love affair with automobiles, but it could also be said that our affair is really with the oil that makes the autos possible, or maybe with the refined oil called gasoline. Yet in many ways our relationship with gasoline is more like an unhappy marriage. Gasoline stinks. It pollutes the atmosphere with carbon dioxide, and it is blamed for global warming and for nitrous oxides that cause smog. It finances totalitarian dictatorships and terrorists, and drives us to war.

We seek a divorce, but the steps seem too difficult or too painful, and what would we be left with, particularly if our standard of living is tied to our energy consumption? Are we really trapped in a doomed marriage? Many people think yes, and the only questions are how long it will last and what kind of terrible consequences we will suffer. After all, isn't oil about to be used up, any decade now? Isn't the skyrocketing price of oil simply a symptom of this looming disaster?

These questions have economic, political, sociological, historical, and psychological dimensions. Some people would add their contention that oil is also driven by national and international conspiracies—or at least by monopolies. But it turns out that the past and future of oil, and of the alternatives, have a very large physics component. These physics dimensions are sometimes more important than the others, so it is essential to know and understand them. And the alternatives to oil also have key physics aspects. I'll bring in engineering and economic aspects as they are needed, but the focus here will be on the science. Laws of countries can be changed, but laws of physics are pretty much set.

The fundamental physics reason for our addiction (or marriage) to oil is the same reason that oil was the weapon of choice for the 9/11 terrorists: it carries huge amounts of energy. Consider the fact that our best rechargeable batteries hold only 1% of the energy of gasoline. Now ask, Why don't we drive electric autos? Were they killed by a conspiracy? Conspiracy or not, it hardly matters. This energy storage discrepancy provides a huge physics barrier. Batteries simply do not store much energy—not when compared to gasoline.[2] We love gasoline because it is so energetic!

Here is another example that illustrates the enormous energy of gasoline: For the same weight, gasoline delivers 720 times the energy of a bullet.[3] Are you surprised by this number? It makes physics sense, if you think about it in the following way. Bullets are driven by the explosion of gunpowder or other similar explosives, but these typically have about 15 times less energy than gasoline. Moreover, they don't put all their energy into the motion of the bullet; much is lost to heat of the expanding gas. (The longer the rifle is, the more energy is transferred to the bullet.) Add to all that the fact that the explosive charge usually weighs much less than the bullet itself, and you should not be too surprised that bullets carry so little energy compared to an equal weight of gasoline.

To provide its energy, gasoline must combine with oxygen. In an auto, the fuel is mixed with air by the fuel injector or carburetor. Part of the energy advantage of gasoline comes from the fact that we don't have to carry oxygen with us; it is as free as the air. Let's compare the chemistry of gasoline and gunpowder. Gasoline consists of atoms of hydrogen and carbon—about two hydrogens for each carbon. When it burns or explodes, the hydrogen combines with oxygen to make water (H_2O), and the carbon combines with oxygen to make carbon dioxide (CO_2). If the combustion is incomplete, gasoline also makes carbon monoxide (CO). In contrast, gunpowder uses potassium nitrate in place of oxygen, as its *oxidizer*, so it doesn't need to pull in any air. That allows the reaction to occur in condensed form and much faster than with gasoline. Rockets also must carry an oxidizer, sometimes in the form of liquid oxygen or hydrogen peroxide.

Gasoline is even better than food. (No, I don't mean the taste, I mean in energy per pound.) Here are some less astonishing but maybe still somewhat surprising facts. For the same weight, gasoline delivers approximately

- 4 times the energy of steak
- 2 times the energy of chocolate chip cookies
- 1.4 times the energy of butter

Perhaps the most surprising result is that food has even this much energy, almost as much as gasoline and much more than TNT. Here is the direct comparison between food and explosive: steak has almost four times the energy of TNT; chocolate chip cookies, eight times as much. No wonder we are addicted to food! If you find this high energy content implausible, watch a hummingbird. It uses enormous energy to flap its wings just to sip a tiny amount of nectar. Clearly the energy in the nectar must be more than enough to cover the work being done by those rapidly beating wings that hold the bird in front of the flower. It is. Food is almost as good as gasoline.

You probably eat between 1 and 2 pounds of food each day. (I know it seems like more, and if you are under age 20 it may be more, but add it up and see. Don't count the water.) Yet all your work, your thinking, your accomplishments are achieved on the energy in that remarkably small amount of food.

On the downside, the enormous energy content of food is what makes it so difficult to lose weight without serious dieting. One 12-ounce can of soda contains, typically, 150 food calories. A person can work that off with a half hour of vigorous exercise (running, not jogging; basketball, not baseball; swimming, not golf), provided, of course, that he doesn't reward himself with a can of soda. The best way to lose weight is to eat less, not to exercise more.[4]

What about other energy sources? Here is a comparison with four important alternative fuels. For the same weight, gasoline delivers approximately

- 2 times the energy of coal
- 2 times the energy of methanol (wood alcohol)
- 1.5 times the energy of ethanol (drinking alcohol)
- 1.1 times the energy of butanol (a likely future biofuel)

These facts are important when you consider alternatives to gasoline. If you live in a state that sells ethanol as a substitute, or perhaps ethanol mixed with gasoline in a combination called *gasohol*, then you've possibly been intrigued by the fact that it costs less per gallon than gasoline. But look at the numbers that I just gave. Per pound and also per gallon, alcohol delivers less energy. In fact, when computed per mile, alcohol sold in the United States is more expensive than gasoline. Of course, some people use alcohol or mixed fuels not because they are cheaper, but because they believe that these fuels are better for the environment. They sacrifice wealth for the good of the world. They may be wrong, however. Ethanol made from corn saves the world very little pollution; we'll discuss this more when we discuss biofuels. Butanol looks attractive because of its high energy density—comparable to that of gasoline—and for this reason butanol may be the biofuel of the future.

Let's wind up this introduction with a few more numbers, several of which are important, and one of which (the last one) is just amusing. The following fuels do *beat* gasoline in energy per pound:

- *Natural gas* is 1.3 times better
- *Hydrogen gas or liquid* is 2.6 times better
- *Uranium or plutonium fission* is 2 million times better
- *Hydrogen fusion* is 6 million times better
- *Antimatter* is 2 billion times better

The 2.6 value for hydrogen gas is what inspires people to talk about the hydrogen economy. With 1 pound of hydrogen fuel, you can go 2.6 times farther than with gasoline. Yet a pound of hydrogen, even in its liquid form, takes up a lot more space. That's why hydrogen has 4.5 times less energy per gallon, as I stated at the beginning of the chapter.

The uranium number may not surprise you; the high value is why it is used in nuclear bombs and nuclear reactors. Fusion is even better, and it uses as its fuel heavy hydrogen, a component of ordinary water—something we will not run out of. (Of course, the same was once believed about wood from forests.)

I included antimatter in this list only because it is so prominent in science fiction. That number will be most useful for future presidents to know when they talk to teenagers. Someday (in a few hundred years?) we might actually use antimatter as fuel, if we can figure out a good way to bottle it; that is a problem, since it explodes when it contacts ordinary matter. Antimatter has an important feature in common with hydrogen gas: Neither one is a source of energy. Rather, they are both means of transporting energy. To use antimatter as a fuel, we first have to create it, and that takes more energy than we get back. To use hydrogen as a fuel, we must first extract it from its natural compounds.

Unlike oil, we can't mine hydrogen gas from the Earth. The hydrogen that is present has all already "burned"—that is, combined with oxygen to make water (H_2O), or with carbon to make sugars, starches, and hydrocarbons (including plant matter, wood, oil, and natural gas). To use hydrogen we have to separate the hydrogen from the other atoms. We can remove the hydrogen from water by running electric current through it—a process called *electrolysis*. But that process takes energy, and when we use the released hydrogen as a fuel, we get back only 30% to 40% of the energy that we put in; the rest is wasted as heat. Beware of inventions that claim to use ordinary water as fuel; these usually obtain the hydrogen by using other energy to separate it from water, by electrolysis or use of another fuel such as a purified metal.

There is one way to get net positive energy from hydrogen: obtain it from natural gas. This is, in fact, the way we get most of our hydrogen today. Natural gas is mostly methane, CH_4, with molecules that consist of one carbon atom and four hydrogen atoms. When methane is reacted with water, out comes hydrogen and carbon monoxide (along with some carbon dioxide). This hydrogen can be used as fuel, but the energy we get is less than we would have obtained directly from the methane.

Power

If TNT contains so little energy, why do we use it at all? The answer is that it delivers a lot of *power*. In popular parlance, power and energy mean the same thing, but scientists like to make a distinction—one that will be useful for future presidents to know: power is the *rate* at which energy is used. Power can be measured in terms of calories per hour or joules[5] per second. TNT has less energy than gasoline, but it delivers what little it has with such speed that it can shatter rock. Gasoline has more energy than TNT, but TNT can deliver more power than gasoline can.

The same amount of energy can be delivered at different rates— that is, at different powers. In the chapter on 9/11 we discussed a hammer as a force multiplier. Now let's take another look, to see that it can also be considered a power multiplier. When you accelerate a hammer, you are putting energy into it over the length of the swing. You do this relatively slowly because your arm has limited power. Just before hitting the nail, the head of the hammer has all the energy that you are going to put into its motion. This energy is called *kinetic energy* because it is stored in the movement of the hammer. When the head finally makes contact, it delivers that energy very quickly to the nail. The energy that the hammer puts out is the same that you gave it with the swing, but because it gives up that energy much faster, we say it has more power. Of course, it delivers the higher power for a shorter time. The greater power means that the hammer will put a larger force on the nail than you put on the handle of the hammer; thus the force is multiplied, and the greater force splits the wood.

James Watt was the first person to measure the power that a healthy horse can deliver, and he called that unit the *horsepower*. The term is used most commonly today to describe the power of the thing that replaced the horse: the automobile. Another unit of power was named after Watt himself, and is called the *watt*. A thousand watts, one *kilowatt*, is approximately one horsepower.[6] That fact is so useful that a future president should memorize it:

1 horsepower = 1 kilowatt

This rough equality will prove very useful in getting a feel for power. Solar power, for example, is about 1 kilowatt per square yard (or square meter, if you prefer). You can now visualize the power in a square yard by thinking of it as the power of a horse. That sounds like a lot, and it is, but it is not enough to run a modern car. Typical cars have engines that can deliver 50 to 200 horsepower.

Watts are usually used to measure electric power. Lightbulbs are labeled by how many watts they use. If you turn on ten 100-watt lightbulbs, you are using $10 \times 100 = 1000$ watts = 1 kilowatt of electric power. That's about one horse, and one square yard of sunlight.

Now for the confusing part: measuring energy. If you use a kilowatt for an hour, you have used an amount of energy that we call a *kilowatt-hour* (abbreviated *kWh*). What makes the term confusing is the presence of the word *watt*. But you should be no more confused by these two words than you are by *miles* and *miles per hour* (*mph*). One is an amount (miles), and the other is a rate (miles per hour). For power and energy, the kilowatt is the rate of energy delivery (the power), and the kilowatt-hour is the total amount of energy delivered. A similar confusion arises from the term *light-year*, which is not a measure of time, but refers to the *distance* that light travels in a year. A light-year is a distance, and a kilowatt-hour is an amount of energy.

The average US household uses about a kilowatt of power. In a 24-hour day, the same household will use 24 kilowatt-hours of energy. Most future presidents are wealthier than average and probably use more. If you use a kilowatt for an hour, the electric utility will charge you for 1 kilowatt-hour of energy. The cost depends on your location, but the average in the United States is about 10 cents for that kilowatt-hour. That is a lot cheaper than renting a horse for an hour! Twenty-four hours of using a kilowatt would cost you $2.40. A 30-day month would cost $72. A year, 365 days, would cost $876. It adds up. And your household probably uses more than a kilowatt.

A thousand houses will use 1000 kilowatts—also called a *megawatt*, meaning a million watts. Intermediate-sized power plants produce 50 to 100 megawatts of electricity, and they often serve local communities. The largest electric power plants produce about a billion watts of electric power, called a *gigawatt*. To avoid using yet another term, however, some energy experts talk about "thousands of megawatts" rather than gigawatts. The total electric power of the United States averages about 450 gigawatts. The electric power used by California amounts to about 40 gigawatts. These are useful numbers to know. Imagine 40 large power plants supplying all of California's power. To remember the value for the United States, remember that the whole country uses just over 10 times that amount: 450 big power plants, each producing one gigawatt.

Here is a conversion to our old energy unit, the food calorie. One watt-hour is equivalent to about one food calorie.[7] In other words, a flashlight turned on for an hour consumes one calorie. If we multiply by 1000, we get the following approximate conversion:

$$1 \text{ kWh} \approx 1000 \text{ food calories}$$

A typical adult consumes 2000 food calories per day. That's equivalent to 2 kilowatt-hours, or about 20 cents of electricity. Are you surprised that electric energy is cheaper than energy from food? Actually, food isn't very expensive if we stick with basics. Enough rice to provide your 2000 calories—1.3 pounds—costs about 70 cents in a US grocery store. If you buy it by the ton, you can buy that 1.3 pounds of rice for less than 10 cents. You can live on about a dime a day of food! It is only the fancier food that is really expensive.

Energy Alternatives

When thinking about alternatives to petroleum and other fossil fuels, it is important to know how the United States uses its current supplies. About 5% is used to manufacture materials such as fertilizer, chemicals, and plastics. The rest is used to generate energy. Here's roughly how that breaks down for the US use of fuel:[8]

- 28% is used for transportation (gasoline and jet fuel).
- 40% is used to generate electric power.
- 20% is used for direct heating (natural gas, coal).
- 32% is used by industry.

These numbers add to more than 100% because of overlap; some of the electric power, for example, is used by industry.

Rather than learning these specific numbers, all you really have to know is that fuel is used in four different ways—transportation, electricity, heat, and industry—all in comparable amounts. That fact has important policy implications for future presidents. For example, if you are concerned that we are polluting the atmosphere by driving automobiles and you decide to solve this problem by replacing all gasoline with biofuels (alcohol made from plants), you will affect only 28% of the total.

Equally important (and interesting) is the broad spectrum of US energy sources:

- 29% from imported oil
- 11% from domestic oil
- 24% from coal
- 19% from natural gas (methane)

8% from nuclear

8% from others (solar, hydro, wind, biomass, geothermal)

Again, the diversity of sources is the most important fact. Here is an example of the way you might use these numbers for an important policy decision: Suppose someone proposes to replace all of our fossil fuel electric power plants with nuclear power, in an effort to reduce CO_2 emissions. The first list tells you that 40% of our energy is used to create electricity. The second list tells you that 8% is already coming from nuclear. So making the change would affect only the remaining 32% of our power use. Remember, if we are going to reduce fossil fuel emissions, we have to address several sectors, not just one.

The Bottom Line: The Cost of Energy

Not all energy sources are equally expensive; in fact, you may find the differences more astonishing than any of the astonishing numbers I've shown so far. Here is the most important fact: for the same energy, coal in the United States is 20 times cheaper than gasoline. That number is important for future presidents to consider. It implies that some developing nations are likely to rely on coal for their energy needs, rather than on oil or natural gas.

Here are some details. The following list compares the cost of energy per kilowatt-hour from various sources. The list does not include the cost of the plant and the power lines that deliver the energy.

- Coal: 0.4–0.8¢ ($40–80 per ton)
- Natural gas: 3.4¢ ($10 per million cubic feet)
- Gasoline: 11¢ ($3.70 per gallon)
- Car battery: 21¢ ($50 per battery to replace)
- Computer battery: $4 ($100 per battery to replace)
- AAA battery: $1,000 ($1.50 per battery)

It is odd that energy cost depends so much on the source. If the marketplace were "efficient," as economists sometimes like to postulate, then all these different fuels would reach a price at which the cost would be the same. This hasn't happened, because the marketplace is *not* efficient. There are large investments in energy infrastructure, and the mode of delivery of the energy is important. We are willing to spend a lot more for energy from a flashlight battery than from a wall plug because the flashlight is portable and convenient. Locomotives once ran on coal, but gasoline delivers more energy per pound, and it does so without leaving behind a residue of ash, so we switched from steam to diesel locomotives. Our automobiles were designed during a period of cheap oil, and we became accustomed to using them as if the price of fuel would never go up. Regions of the world with high gas prices (such as the countries of Europe) typically have more public transportation. The United States has suburbs—a luxury that is affordable when gas is cheap. Much of our way of living has been designed around cheap gasoline. The price we are willing to pay for fuel depends not only on the energy that it delivers, but also on its convenience.

The real challenge for alternative energy sources is to be more economically viable than coal. When we talk about global warming (in Part V), we'll discuss how coal is one of the worst carbon dioxide polluters that we use. To reduce our use of coal, we could, of course, tax it. But doing that solely in the developed nations would not accomplish much, since the ultimate problem will be energy use by nations such as China and India. Leaders of such countries might choose to get their energy in the cheapest possible way so that they can devote their resources to improving the nutrition, health, education, and overall economic well-being of their people.

6

SOLAR POWER

SOME experts say that solar power has no future. They claim that useful power sources must be compact, as if reciting a law of physics. But there is no such physics principle, and even people who appear to be experts can have their numbers very wrong. Let me tell you a true story that illustrates the kind of confusion that reigns over solar power.

An Anecdote

Liz, a former student of my class in Berkeley, came to my office. She was eager to share an experience that she had had a few days earlier. Her family had invited a physicist over for dinner, someone who worked at the Lawrence Livermore National Laboratory. He regaled the family throughout dinner with his stories of controlled thermonuclear fusion, its use for national security testing, and its great future for the power needs of our country. According to Liz, the family sat in awe of this great man describing his great work. Liz knew more about fusion than did her parents, because we had covered it in our class, but still not very much. She said she learned a lot.

After a period of quiet admiration at the end, finally Liz spoke up. "Solar power has a future too," she said.

"Ha!" the physicist laughed. (This is as related to me by Liz. He didn't mean to be patronizing, but this is a common tone that some physicists affect.) "If you wanted enough power just for California," he continued, "you'd have to plaster the whole state with solar cells!"

Liz told me that she answered right back. "No, you're wrong," she said. "There is a gigawatt in a square kilometer of sunlight, and that's about the same as a nuclear power plant."

Stunned silence followed. Liz said he frowned. Finally he said, "Hmm. Your numbers don't sound wrong. Of course, present solar cells are only 15% efficient . . . but that's not a huge factor. Hmm." He then said he would rethink the issue.

Yes! I have never been prouder of a student. Liz did exactly what a future president needs to be able to do. Not integrals, not roller-coaster calculations, not pontifications on the scientific method or the deep meaning of quantum physics. What she did was far more important: she was able to shut up an arrogant physicist who hadn't done his homework. Liz hadn't just memorized facts. She knew enough about the subject of energy that she could confidently present her case under duress when confronted by a supposed expert. She remembered the important numbers because she had found them fascinating and important. They had become part of her, a part she could bring out and use when she needed them, even a year later.

As a future president, you should aspire to no less.

Basic Facts

With a few basic numbers memorized, a future president can think more clearly about the potential of solar power. We discussed the power in sunlight earlier; now let's look into it a little more carefully. Overhead sunlight delivers to the ground approximately

- 100 watts per square foot
- 1 kilowatt per square yard (or square meter)
- 1 horsepower per square yard (or square meter)
- 1 gigawatt per square kilometer
- 3 gigawatts per square mile

Consider the fact that sunlight delivers a horsepower in just a square yard. Think about that. Visualize it. Does that sound like too much? If it does, try thinking about the intensity of sunlight. Sunlight dries clothes much faster and far more thoroughly than a household clothes dryer. Every cold camper appreciates the morning appearance of the sun. Watch a sunrise and experience the sudden warming of your face once the sun peeks above the horizon. Sunlight delivers enormous power density. No wonder many early peoples worshiped the sun!

A kilowatt per square yard is large, but in another sense it is remarkably small. Let's consider a solar-powered auto. Put a large solar cell on the roof of the car. A reasonably expensive solar cell can convert about 15% of the solar power into electric power. So if the cell has an area of 1 square yard, it will deliver about 15% of a horsepower—that is, about $\frac{1}{7}$ horsepower. That's about the same as the power that a healthy person can produce by pedaling a bicycle.

Little wimpy cars, like the 1966 Volkswagen Beetle that I owned for many years, deliver a maximum of 50 horsepower. That's 350 times as much as we could get from the square-yard solar cell. Big muscle cars deliver 200 horsepower. That's 1400 times as much as we get from our solar cell. So, although solar power is intense, a whole horse in just 1 square yard, it is feeble compared to the power we need to drive an auto. Here's the bottom line: solar autos will never be a realistic technology to replace our gasoline-driven cars. This conclusion is a consequence of physics, with only a little bit of sociology: my assumption that you will not be satisfied with a car that delivers no more power than you could pedal.

Could solar cars become practical if solar cells improve? The best solar cells tested in the laboratory today have an efficiency of about 41% and cost about $100,000 per square yard. Let's be optimistic and assume that we can one day actually get 100% efficiency and that the solar cell will be cheap. Even so, we will get only 1 horse-power for our car, regardless of the price. Double the size of the solar cells to 2 square yards, and we'll get 2 horsepower. There's no room to put up more solar cells.

Piddling horsepower doesn't mean that solar autos don't exist. In fact, several races every year are dedicated to such vehicles. Some of them use expensive solar cells, with efficiency of 30%—double the cheap cell number. But all the autos are low and sleek to avoid air resistance, because it is hard to go very fast with only a fraction of a horsepower. To go up a hill in such cars requires accumulating energy on the flats by charging a battery with whatever excess power can be spared. I repeat: solar autos are not in our future.

An average US house uses 1 kilowatt, about 1 horsepower. Could a house be powered by a solar cell? Solar cells are typically 15% effi-cient. Moreover, the sun is not always out, and it is rarely overhead. Combining these factors shows that the solar cell average efficiency is only a few percent. So you could do it, if you had, say, 20 square yards of solar cells. That area of cells will fit on the roof of many homes. Some people are already doing it. It is environmentally very clean, and it sounds very cheap.

Why doesn't everyone put solar cells on the roof? Let's consider the cost. As of 2008, the typical expense (in sunny California) is $3.50 per installed noontime watt. Average in the time when the sun isn't overhead, and that amount becomes $14 per installed watt, $14,000 for your one-kilowatt home. That still sounds pretty good. You invest $14,000, and you don't have to pay the utility. How much do you save? Energy from the electric power companies averages about 10 cents for 1 kilowatt for 1 hour. Since there are 8760 hours in a year,[9] the power company's typical charge amounts to $876 per year. That is how much you would save if you installed the solar cells on your roof. In effect, you are *earning* $876 per year on an investment of

$14,000. That works out to a 6.2% average return on the invest‑ment—better than you can do with a typical savings account. So you are making money, as long as you don't have to replace the cells.

Suppose, on the other hand, that the cells last only 10 years. Then there is an additional replacement cost that averages $1,400 per year. Instead of making a yearly profit of $876, when you subtract the cost of replacement you suffer a net $524 yearly loss. To break even, an actuarial calculation[10] (assuming a 3% interest rate) shows that the cells would have to last 22 years. If they last that long, then you are just breaking even by investing in solar cells. If they require replacement or repair sooner than that, then you are losing money. If they last longer, then you are coming out ahead.

Right now, these are the economic and physics reasons that are delaying our transition to solar power. People who are wealthy enough can convert, and feel good about the societal good they are accomplishing. What is the prognosis for more widespread use? Improving efficiency might help, but that will be only a small factor. The real potential comes from a possible substantial reduction in the cost of the cells.

Part of the problem with solar power is that it is so inefficient to turn it into electricity—only 15% for inexpensive cells. If you don't need to convert it, however, but just use the heat directly, then it becomes much more attractive. For heat it is effectively 100% effi‑cient,[11] and it doesn't require expensive solar cells. For this reason, solar-powered water heaters make good economic sense in many parts of the country. But when you add in the cost of the plumbing, do you really save money? That depends on the alternatives. Let's consider a few. A ton of coal costs only $40, and it will supply a kilo‑watt of heat for most of a year. Electric heating is the most expen‑sive, but for short periods even electric costs are tolerable. For example, you can warm your bedroom on a cold night with a kilo‑watt electric heater. At 10 cents per hour, that will cost less than a dollar for the night. That sounds like a bargain, but it does add up. A dollar per night is $365 per year.

Solar Power Plants

So far we have been talking about using solar power for personal use—automobiles and houses. What about large power plants? There are two main approaches to using solar power to generate electricity. The first is to use solar cells similar to the ones we've been discussing, but to deploy them in huge numbers over large areas. The second method is called *solar thermal conversion*.

Let's begin with the solar cell approach. The key material in these cells is silicon, a very abundant element on Earth and the heart of our semiconductor industry. Most of our modern computer electronics are based on the electric properties of silicon, and that's what gave rise to the name *Silicon Valley*. When sunlight hits silicon, it knocks an electron out of the silicon crystal and transfers at most 25% of its energy to the electron. (The rest goes into heat.) That electron can then be drained off to supply electric current. Silicon solar cells are now in widespread use. As I said earlier, inexpensive solar cells aren't as efficient as the best ones, and achieve only 10% to 15% energy conversion to electricity.

More efficient solar cells are possible, but making them requires complex manufacturing methods and expensive materials. The trick is to make multiple layers, one for each different color in the spectrum of sunlight. The best efficiency achieved so far, 41%, has been with a "triple junction" cell[12] that has separate layers for different colors of light. But they are expensive, about $65 per square inch. If the cost can be brought down, and if we don't run out of the specialized materials needed to make them,[13] then solar could become an important source of energy in the future. I'll come back to this issue when I discuss global warming and alternative energy in Part V.

The second method to produce electricity from sunlight, solar thermal conversion, begins with reflectors or lenses to concentrate the sunlight. Just as a large magnifying glass can be used to start a fire, this focused sunlight can be used to boil water. The resulting steam then runs a steam engine that generates electricity. Figure 6.1 shows a solar thermal conversion plant situated near Seville, Spain. It produces 11 megawatts of electric power, enough for 60,000 people.[14]

Although the rays in the photo look like they are emanating from the tower, in fact they are reflecting off 624 large mirrors on the ground, each one more than 1000 square feet in size. The sunlight is

Figure 6.1. Solar power plant near Seville, Spain. The mirrors focus sunlight onto a boiler, and the resulting steam runs a turbine that produces electricity.

directed onto a central boiler that sits on the top of the 377-foot tower (taller than the Statue of Liberty). The sunlight heats the water to 750°F. The whole system is reasonably efficient, converting about 15% of the intercepted solar power to electricity. There are plans to expand the plant to 300 megawatts in the near future.

The Seville solar power plant seems to set an example for the rest of the world to follow. Why aren't we doing this? What's the catch? Is there an anti-solar conspiracy? No. It's the same old catch: cost. The electricity from the plant costs about 28 cents per kilowatt-hour (versus 10 cents from fossil fuels). This uneconomical plant was made possible by the Spanish Royal Decree 436/2004, which subsidizes the operation. The Spanish government did this to help meet its requirements under the Kyoto Protocol limiting CO_2 emissions. (I'll discuss Kyoto further in Part V, which covers global warming.) It was also hoped that building the technology will eventually reveal ways to save money.

Solar Airplanes

Solar cars, with less than 1 horsepower, are just for hobbyists. It is therefore somewhat surprising to discover that solar airplanes have real value, serving some very practical applications. Their main use in the future will probably be for continuous spying on critical areas of conflict.

Solar aircraft (drones, actually, with no passengers) have already flown. One of the most successful was called *Helios*, shown in Figure 6.2. The solar cells are on the upper and lower surfaces of the wings; the cells on the undersides use light reflected off the Earth. The *Helios* vehicle carried fuel cells, a kind of rechargeable battery, that were charged during the day to allow continued flight at night. The payload was 100 pounds, enough for a good telescope, camera, and radio. In 2001, *Helios* achieved an amazing altitude of 96,863 feet (commercial airliners fly at about 40,000 feet). It was built by AeroVironment, a company started by engineer Paul McCready, who designed the human-powered airplanes the *Gossamer Condor* and the *Gossamer Albatross*.

Many follow-up solar-powered planes are now under test or development. The most prominent of these is called *Pathfinder*. It is somewhat smaller than *Helios*, and designed for shorter missions. The maximum power delivered by its solar cells is 17 horsepower. It flies at the leisurely speed of 20 miles per hour.

Imagine that you are president and you need a surveillance system above a particular country, perhaps North Korea or Iraq. You want to send a camera there that can circle above one location, maybe a nuclear test site, to watch for suspicious activity. You want it to stay there, perhaps for months. Your secretary of defense says that's impossible; no craft can carry enough fuel. You ask, "What about solar energy?" A solar plane can get through the night by stored battery energy, or just by gliding. It flies so high that it is well beyond the range of manned aircraft. In fact, with its low velocity, it doesn't really stand out in a radar scan. It's possible that no one will even notice that it is there. The secretary of defense asks whether it could carry the camera and radio equipment needed for such a mission. You reply, "The *Helios* mission had a payload of 100 pounds. See if your scientists and engineers can squeeze what they need into such a package."

Figure 6.2. Helios *in flight. In this image, the curvature of the wings is exaggerated by the fact that it was taken from an angle with a telephoto lens. The wingspan is 247 feet, 36 feet longer than that of a 747.*

7

THE END OF OIL

MANY people believe that we are soon going to run out of oil. Actually, according to a famous analysis by Marion Hubbert done in 1956,[15] we will never completely run out. The plot in Figure 7.1 shows the historical oil production up to 2004, and then the expected production until the year 2050. Although the amount of oil recovered each year will go down, it drops by only a factor of two by 2050.

Although the general shape of the curve was not widely antici-

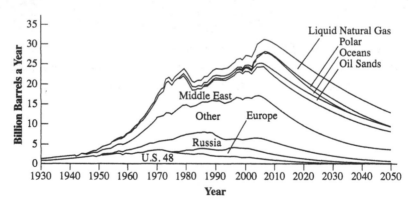

Figure 7.1. The Hubbert curve for oil. The plot shows the growing consumption of oil in the past, the peak (occurring about now), and the predicted drop in usage as world supplies are depleted.[16]

pated prior to Hubbert's paper, the reason for the shape now seems obvious. When oil was first used as a commodity, it was rare and expensive, and only little was produced. Thus, people were encouraged to search for more, and as they succeeded the production rose; that's the left side of the plot. The world is finite, however, so after a few decades little additional oil was being found. (The North Slope of Alaska, one of the more recently discovered sources, produced only 2 billion barrels of oil in its maximum year, and it yields less than 1 billion barrels per year now. That's only 3% of the total world production of 30 billion barrels per year.) For that reason, the production peaks and eventually drops. That happened for US oil in the mid 1970s, and it should happen for the world total productivity soon—maybe as soon as 2008. The maximum is called the *Hubbert Peak.* You need to know that term. You'll hear politicians (your rivals) using it to show their erudition. It is becoming so widely used in popular analysis that an episode of the TV series *The West Wing* was named after it.

Notice how slowly the Hubbert curve drops. Hubbert did not predict the slowness of that drop, although he was careful to state his limited assumptions. Part of the reason for the slow decline is that as time passes and the price of oil remains high, new technologies emerge that enable more oil to be recovered. Moreover, oil usage can fluctuate, as it did in the 1970s because of the oil embargo, suddenly dropping and then rising again.

Underground oil is not found in caves or large openings, but in microscopic cavities in porous rock. When the oil is first drilled, it tends to come up of its own accord, from the enormous pressure as the weight of the material above squeezes the saturated rock. Only 20% to 30% comes up that way, however. To get the rest, gas such as carbon dioxide or liquid such as water is pumped down into the rock. (Pumping carbon dioxide into these wells will prove to be a useful means to keep it out of the atmosphere; we'll describe this further when we discuss global warming.) With these methods, an additional 30% to 60% of the oil can be pushed out. What about the rest? Getting it out is difficult because the oil tends to stick to the rock. All sorts of ideas have been tried to overcome this problem—from pumping detergents down, to injecting polymers that make the oil less viscous. It has even been suggested that bacterial biofilms could be grown at the oil–rock interface to make the surface slippery.

Hubbert originally thought that the world production of oil would peak in the mid 1990s, about a decade earlier than seems to be happening. The delay was largely a result of the development of new technologies and the willingness of the world to buy oil at above $100 a barrel. In 1956, when Hubbert wrote his paper, the total recoverable oil in the world was thought to be less than a billion barrels. That much has already been pumped out of the ground. With enhanced oil recovery, and the possibility of recovering oil from oil shales and oil sands, the total recoverable reserves now appear to be an additional 5 billion barrels.

Because of the limited pumping capacity of the world's oil wells, demand has now outstripped the ability of the existing wells to keep up. In particular, the rapid growth of the economies of China and India has sopped up most of the available oil-pumping capacity. Competition among buyers pushes the price higher. Oil drilled in Saudi Arabia for $2 per barrel is sold to the United States for $100 per barrel and higher.

In the past, the OPEC cartel did not let prices rise high. OPEC's public reason for this control was that it wanted the Western economies to stay vibrant. However, most experts think that OPEC's motive was more self-serving. Once the price of oil reaches $50 per barrel, there are many alternatives. In the 1970s, OPEC's primary competition was energy conservation; that led to the drop in oil prices in the early 1980s. But ultimately, the competition that OPEC fears most is coal.

Oil From Coal: Fischer–Tropsch

Coal is cheap and—unfortunately for OPEC—abundant in the countries that need the most energy: the United States, China,

India, and Russia. Once the price of oil rises above about $50 per barrel, these countries can take their coal reserves and convert them to oil, using a series of chemical reactions known as the *Fischer–Tropsch process*. The basic method is to combine the carbon with the hydrogen from water to make hydrocarbons—the basic molecules of oil.[17] This process was used by Nazi Germany during World War II, when the Germans could not get oil because of the Allied blockade. It was used by South Africa during the era of apartheid for similar reasons, and because the plants still exist, South Africa is still converting coal to oil today. The only reason that we are not yet doing this in the United States is that building Fischer–Tropsch plants is expensive, and nobody wants to make the investment unless it is certain that the price of oil will stay high. Several companies, in fact, want to go ahead and build such plants, but they fear they cannot take the risk unless the US government guarantees that it will cover their losses if oil prices drop again.

Our ability to manufacture oil from coal means that as long as we can afford a price as high as $50 a barrel, we are not going to run out of liquid fuel at that price, not for centuries—at least not once we have our Fischer–Tropsch plants built. The Hubbert Peak for oil does not take into account the availability of the Fischer–Tropsch method. If we are driven to other fuels, it will not be because we can't get oil. Rather, it will happen for one of two reasons: either the alternatives will become cheaper than oil, or we will be forced to use them anyway because of the environmental consequences of the continued use of fossil fuel.

How soon will we run out of coal? The United States has enormous reserves. About 2 trillion tons of coal are known, but twice as much coal might be present. We consume about a billion tons per year. If our coal use doesn't grow, the supply will last over 1000 years. If we wind up using a great deal more—to replace expensive oil, for example—then it might last only several hundred years. Of course, by then we may be using fusion power or solar.

The Price of Oil

To discuss the real price of oil (versus the perceived price), we have to start with a little economics. Economists try to account for inflation so that they can compare prices in *constant dollars*. The plot in Figure 7.2 shows the price of oil, in dollars per 42-gallon barrel, from 1970 to 2007. Note that the high price of oil, over $100 per barrel in 2008, is not unprecedented. The cost approached similar levels during the Iran crisis in the early 1980s. Prior to 1970, the price had been less than $20 per barrel for decades.

There is much more that a future president needs to know about energy. Part III will be devoted entirely to one key field: nuclear energy. Energy will also be at the center of our discussion of global warming in Part V.

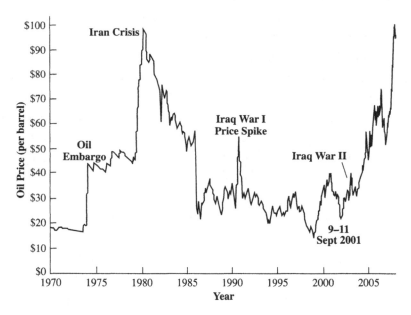

Figure 7.2. Oil prices (adjusted for inflation) from 1970 to the present.[18] *Notice that the current price for 2008, when measured in constant dollars, is only a bit higher than the price in the early 1980s.*

ENERGY: PRESIDENTIAL SUMMARY

IN DEALING with energy, you will be confronted with the fact that the public has a vast amount of misinformation. Many of the facts I gave in the previous three chapters are directly useful for policy decisions, whether they relate to battery-driven autos or energy storage with hydrogen, but equally important are the facts that most people think are true but aren't. Our love affair with fossil fuels ultimately derives from the fact that they are so cheap.

It is important to realize that we are not running out of fossil fuels—but only out of oil. Our coal will last centuries. That is both the good news and (from a global-warming perspective) the bad news. We'll defer our in-depth discussion of global warming to the end of this book. Coal can be converted to gasoline, but we don't presently have the factories to do that. If you want such factories to be built, you may have to guarantee their profit, because otherwise investors will be wary of OPEC's ability to drop prices (as in an old-fashioned gasoline price war) just long enough to put Fischer–Tropsch plants out of business.

There are many alternatives to fossil fuel. We've discussed some of them, and we'll be discussing more in the coming chapters. The key issue will be cost. We have been having a love affair with gasoline not because it smells good, but because it has been very cheap and abundant. Shifting to coal will be expensive (at first) and potentially dangerous, especially from the environmental point of view. There are no simple, obvious solutions, despite hype about the hydrogen economy and the imminent expansion of solar. Coal is the front-runner precisely because it is so cheap. From the environmental point of view, coal is perhaps the worst source of energy. Even if the United States votes in favor of the environment, however, it is not obvious that the developing countries—especially China and India, but also Russia—will go along. That is the conflict you will have to resolve. Only one approach satisfies everyone: conservation. A deeper discussion of that approach will have to wait for Part V.

III

NUKES

THIS *book is radioactive. And so are you. Unless you are dead, in which case we can tell how long ago you died by how much of your radioactivity is left. That's what radiocarbon dating is—the measurement of the reduction of radioactivity of old bones to deduce the time of death.*

Alcohol is radioactive too—at least the kind we drink. Rubbing alcohol usually isn't, unless it was made organically—that is, from wood. In fact, the US Bureau of Alcohol, Tobacco, Firearms and Explosives tests wine, gin, whiskey, and vodka for radioactivity. A fifth of whiskey must emit at least 400 beta rays every minute or the drink is considered unfit for human consumption.

Biofuels are radioactive. Fossil fuels are not.

Of those killed by the Hiroshima atomic bomb, the best estimate is that fewer than 2% died of radiation-induced cancer.

These statements are all true. They are not even disputed, at least by experts. Yet they surprise most people. Once you understand radioactivity, the explanation for these facts becomes obvious. The initial surprise reflects the confusion and misinformation that many people have about radioactivity. But a future president can't afford to be ignorant in this field.

There are likewise many misconceptions about nukes—nuclear bombs and nuclear reactors. Despite the widespread fear, a nuclear reactor cannot blow up like an atomic bomb—even if a team of PhD physicists has had full access to it and tried to make it happen. But then, what really did happen at Chernobyl? Hasn't it been shown that a high school student could design a nuclear bomb? What are the real dangers of nuclear waste? Do we really have to store it for tens of thousands of years? What would it take for a terrorist group, or a "rogue nation," to develop nukes?

A good place to start our study of nukes is with a look at radioactivity—and the threats it poses to life.

8

RADIOACTIVITY AND DEATH

PEOPLE have an almost primal fear of radioactivity. It may be a new example of an old Jungian archetype: the fear of unseen danger, perhaps originally a predator or enemy lurking in ambush. Other incarnations include the fear of witches, germs, communists, or monsters under the bed. But radioactivity is worse. Not only is the threat hidden, but so is the attack. Genes are invisibly mutated, showing no sign of the assault until a decade or two later when the damage finally reveals itself in a growing cancer.

Radioactivity is the explosion of the nucleus, the tiny core of the atom that contains most of its mass and most of its energy. The explosion takes place suddenly and randomly, and releases a million times more energy per atom than TNT; for fission the factor is even greater: 20 million. It's that huge energy that makes radioactivity so dangerous. The most famous radioactive atoms have familiar names: uranium, plutonium, radiocarbon, strontium-90, and (more recently) polonium-210, the material used to assassinate Alexander Litvinenko, the former Russian counterintelligence officer.

Radiation is the name for the fragments that fly out when the nucleus explodes. Just as with a hand grenade, it's the fragments that do the damage. They are the shrapnel, the debris, the bullets. They spew out with enormous velocity, sometimes approaching or equaling the velocity of light.[1] When these chunks rip into your body, they tear your molecules apart. If enough cells are killed, you will die quickly. Lesser levels of damage can still induce cancer.

We measure the damage done to the body in a unit called the *rem*.[2] If you get a whole-body dose of 100 rem, you probably won't notice. Your body will repair most of the damage without even making you sick.

Larger doses are worse. If you are hit with 200 rem, you'll get sick. Your illness is called *radiation poisoning* or *radiation illness*. Most of your hair will fall out, and you'll feel nauseated and listless. You've seen this syndrome if you've known someone undergoing radiation therapy, usually to kill a cancer. The nausea is due, in part, to the fact that your system is working so hard to fix the extensive damage that it cuts back on other energy-intensive activities, such as digestion.

At doses above 200 rem you have a good chance of dying. At 300 rem, your chance of death from the exposure reaches 50%, unless you get blood transfusions and other intense medical treatment. For that reason, the dose of 300 rem is called the *LD50* (for *lethal dose 50%*). Exposure to 1000 rem will incapacitate you within hours. At such a high level, even medical treatment is unlikely to save you.

Future presidents can remember it in this way: *hundreds of rem are really bad.*

Yet a dose below 100 rem causes no symptoms whatsoever. For that reason, we say there is a "threshold" effect. The threshold is sharp because your body is robust. It finds ways to work around the partial damage, and even repair most of it without impairing the normal functioning of your body. Low doses aren't noticed, although they might eventually give you an increased risk of cancer, many years later; we'll get to that in a moment. Most diseases have a similar threshold effect. If your body can recognize a virus and kill it before it multiplies, you won't catch that cold.

The existence of the radiation threshold is very important for future presidents to understand. Suppose you are worrying about a potential dirty bomb (a radiological weapon)—a terrorist attack based on the spread of radioactivity. We discussed this in the chapter on terrorist nukes. You need to know that if the radioactivity is spread widely, then the dose to any person exposed will drop below the threshold. That means that if it is dispersed in this way, there will be no dead bodies on the scene; in fact, maybe no one will even feel sick. For some terrorists, that makes the idea of a dirty bomb unattractive. It is possibly the reason why al-Qaeda told its US operative José Padilla to abandon his quest to build a dirty bomb, and to concentrate instead on blowing up apartment buildings with natural gas. Conventional explosions yield more dead bodies, and that may be what terrorists want for the news stories.

Some cancer victims avoid radiation therapy because they fear the radiation. That is poor reasoning; it is their cancer cells that should fear the radioactivity. Such cells are even more vulnerable to radiation than are normal cells, probably because they are putting all their metabolic energy into unlimited growth rather than into repairing damage. That's why one of the most effective treatments of cancer is to give the victim as much radiation as he can stomach.

How much radiation does it take to give you a dose of one rem? Let's consider gamma rays, a kind of radiation that is very penetrating and the source of much of the potential death from nuclear weapons, fallout, dirty bombs, and reactor accidents. You can think of a gamma ray as an extremely energetic X-ray. To get 1 rem of radiation, your body must be exposed to approximately 10 trillion gamma rays.[3] That's 10,000,000,000,000. It sounds like a large number, and it is. It may horrify you even more to realize that the firefighters of Chernobyl were exposed to hundreds of times more gamma rays than that, over a quadrillion gammas each. That's why they died of radiation illness.

In the next chapter we'll talk about the natural radioactivity found in the human body. Just from carbon alone, our bodies undergo about 120,000 nuclear decays every minute, with each decay emitting a high-energy electron (called a *beta ray*) that damages nearby cells. But even a year of this self-exposure results in less than $\frac{1}{1000}$ of a rem. It is important to realize that although truly intense radiation is deadly, low levels are part of the natural environment. Excluding all man-made sources, the typical American is exposed to about 0.2 rem per year from nature.

Radiation and Cancer

When you are developing in the womb and later as you are growing up, your cells rapidly split to form new cells. Each one is nourished, grows to full size, and then splits again. When you reach adulthood, a series of special genes in your DNA tells the cells to stop, so you no longer grow. The division is restarted only for special purposes—for example, if you need to heal a wound or replace lost blood. When the work is done, the regulatory genes again tell the cells to stop dividing. Because stopping cell multiplication is so important, several redundant genes do it. If you are exposed to radiation and are particularly unlucky, all these regulatory genes may be destroyed. If that happens, the cells return to full-scale division and growth, but with nothing to stop them. Ten doublings, and one cell has become a thousand cells.[4] Ten more doublings, and you have a million cells. Another ten doublings, and you have a billion. This exponential explosion of unwanted cells is what we call cancer.

When a gene is damaged, the cancer may develop immediately, but more likely it will take years or decades before it starts. That's because many cancers require two or more mutations before the cell loses all control over rapid multiplication. If the radiation caused the final mutation needed, the coup de grâce, then the cancer will occur quickly. If the radiation came early (for example, I had X-rays taken of my feet at a shoe store when I was a child, possibly giving me several rem), then the cancer won't develop until another mutation takes place, maybe decades later. If a smoker is exposed to radiation, the dangers of that exposure are higher than for the nonsmoker because the smoker already has many mutations.

Now get ready for a paradox: the dose of radiation needed to cause cancer has been measured to be approximately 2500 rem. That's enough to kill someone within hours of radiation poisoning. Given one cancer dose of radiation, a person won't live long enough to get cancer.

But . . . but . . . if the cancer dose is that high, how can radiation ever cause cancer? The solution to the paradox lies in the theory of chance. Given a small dose—say, 25 rem, equal to 1% of the cancer dose—a person will not get radiation illness, since that dose is well below the radiation illness threshold. But even at that low dose, there is some probability that one of the regulatory genes that controls unlimited cell growth will be damaged by radiation and not be repaired. As a consequence, 1% of a cancer dose will have a 1% chance of triggering an eventual cancer. For larger doses, it is proportional. Four percent of a cancer dose has a 4% chance of triggering a cancer. But there is a limit to this proportionality: people exposed to 100% of a cancer dose of radiation, never get cancer, because they die of radiation illness within hours.

The kinds of measurements that led to this conclusion are summarized in Figure 8.1, a plot adapted from the US National Academy of Sciences report titled "The Biological Effects of Ionizing Radia-

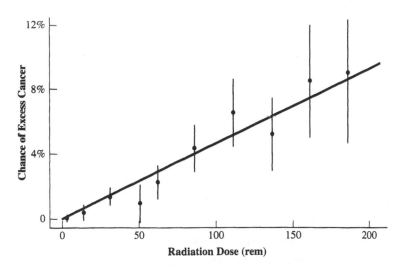

Figure 8.1. Cancer from radiation: the linear effect.

tion," or BEIR report.[5] The plot plays such a key role in the discussion of the dangers of radiation that it is worth understanding.

The horizontal scale shows the radiation dose, in rem, and the vertical scale is the chance of cancer due to radiation. The dots indicate actual rates of excess cancers seen in victims exposed to these doses of radiation. The vertical lines going through these points are *error bars* showing the uncertainty. They are quite long, meaning that there is a lot of uncertainty. Look at the last point on the right. That point is plotted for a dose just under 200 rem. The bar extends all the way from 4% chance of cancer up to 12%. Does that mean we don't know if 200 rem causes 4% or 12% cancer? Not quite; other studies improve the accuracy a bit. But the number is not well known, and that by itself leads to disputes over the best way to calculate expected deaths from radiation exposure. Some people argue that we should be conservative and use the highest values. Others argue that doing so enormously increases the cost for dubious benefit.

The reason that the uncertainty is so large is that many cancers occur even in people not exposed to radiation, and those cancers have to be subtracted. The diagonal straight line goes through most of the error bars, so we say that the relationship is *linear*. That line is what most people remember, and it has become the standard used for determining public policy.

The Linear Effect

Look at the line on the plot in Figure 8.1. The value at 100 rem is 4%. If we plotted the line all the way up to 2500 rem (far off the right side of the chart), it would pass through 100%. That's why 2500 rem is considered to be one cancer dose. Some people argue that the value is somewhat lower, some say higher (remember the uncertainties), but 2500 is a good average, widely accepted figure—a number that future presidents need to know.

Note that the vertical axis is the chance of *excess* cancers. Why do I say excess? The reason is that we already have about a 20% chance of dying of cancer, even if exposed to no human-created radiation. Nobody knows why. The 20% doesn't come from any proven environmental effect. Environmental radioactivity isn't sufficient. We know of no pollutants whose effect should add up to 20%. Where does this come from? It is a scientific and medical mystery.

Here is one possibility: as our cells grow older, some of their genes just fail from "old age." The eminent biochemist Bruce Ames has suggested that some of these cancers may arise from long-term exposure to oxygen, the same chemical that makes fats rancid. If all of the growth regulators fail, even if just for one cell in the body, that cell will then begin to divide out of control, and cancer results. It's not easy to give up oxygen, but this theory has led some people to consume more antioxidants to minimize the damage.

The point is worth repeating. About 20% of us die of cancer from unknown causes. When radioactivity causes cancer, the risk from radiation is added to that basic number. Radioactivity causes *excess* cancers.

I'll put this in perspective with a specific example. Suppose that you are exposed to 100 rem of radiation. That is 4% of a cancer dose, so your chance of dying from cancer has increased from the "normal" 20% to 24%. (The normal rate of 20% varies among different populations, but to make my examples clear, I'll assume that the value is precisely 20% for everyone.)

Now let's use the linear effect to estimate cancer deaths from the bombing of Hiroshima and Nagasaki.[6] About 100,000 survivors were exposed to significant levels of radiation. It is hard to know what dose they got, and a lot of work has gone into estimating the amount. The best estimate is that these survivors received an average dose of 20 rem.

From Figure 8.1 you can determine that the probability of getting an additional cancer from a 20-rem exposure is 0.8%.[7] Of the 100,000 survivors, then, we would expect 800 people to develop and die from an excess cancer. Compare that to the number of people killed from other effects (blast, fire, radiation illness), estimated to be between 50,000 and 150,000. These numbers imply that radiation-induced cancers account for only 0.5% to 1.5% of the total deaths in Hiroshima and Nagasaki. That was one of the surprising facts listed at the start of this chapter. The real reason that so few died of cancer from the bombings is not that radiation is safe. Rather, it is that almost everyone who got a big radiation dose died of other causes.

Remember that, of the 100,000 survivors, we would expect about 20,000 to die of natural cancer. If, for the purposes of illustration, we take our numbers to be exact, then the radiation exposure will increase the expected cancers from 20,000 to 20,800.

There is a psychological effect that all doctors know, and future presidents need to understand. Many victims of a disease feel confident that they know the cause of their illness. Many, perhaps most, of the 20,000 Hiroshima and Nagasaki survivors who contracted cancer from other causes—who would have died of cancer even without the nuclear explosion—incorrectly assumed that their cancers resulted from the bomb. Part of the reason for this misconception is the misguided public perception that cancer is a much rarer disease than it really is. People who claim to know the cause of their cancers usually go unchallenged, even when they are wrong. You don't argue with someone who is dying.

Let's now talk about a relatively recent event that may lead to far more cancer deaths than did the nuclear bombs of Hiroshima and Nagasaki.

The Chernobyl Nuclear Reactor Disaster

In 1986 a nuclear power plant near the town of Chernobyl in Ukraine had a violent accident. The chain reaction deep in the reactor went out of control, a lot of energy was released, and the core of the reactor exploded. It was a small explosion; nuclear power reactors can't blow up like an atomic bomb, for reasons that we'll discuss. Nevertheless, the explosion was big enough to damage the reactor and start a severe fire. A huge amount of radioactivity was released into the atmosphere, perhaps 30% or more of all the radioactivity in the reactor. Dozens of the firefighters who put out the blaze died from radiation sickness.

This was one of the biggest news items of the 1980s; everybody who was an adult at that time remembers it. Radioactivity from the plant was carried by wind over populated areas. Some even reached the United States.

I'm going to give many of the details about this accident, because it is frequently referred to by people hoping to influence policy. The facts can easily be exaggerated or downplayed, so it is useful to know them.

Most of the harm from the accident was done in the first few weeks. Each nucleus can explode only once, so the radioactivity gets used up. After 15 minutes, the radioactivity had dropped to one-quarter of its initial value; after a day, to one-fifteenth; after 3 months, to less than 1%.[8] But there is still some left, even today. Much of the radiation literally went up in smoke, and only the radiation near the ground affected the population. It is hard to estimate the total radiation exposure to humans. About 30,000 people near the reactor are thought to have received about 45 rem each, on average, similar to the average dose received by the survivors of Hiroshima. Note that this average level is too small to induce radiation illness, but that the chance of additional cancer for people exposed is 45/2500 = 1.8%. That risk should lead to about 500 cancer deaths in addition to the 6000 normal cancers from natural causes. The government decided to evacuate all regions in which a person would receive a lifetime dose of 35 rem or more. For most of the region, the radioactivity has now (in 2008) died down to the level well below 1 rem per year, so in principle, people could now move back.

Here's a tough question of the kind that a president might have to address: Was the evacuation of the Chernobyl region wise? To think about this, imagine that, instead of being a future US president, you were a resident of Chernobyl right after the accident. Imagine that unless you left, you would get a radiation dose of 45 rem. As I have just shown, that dose increases your cancer risk from 20% to 21.8%. If given the choice, would you give up your home in order to avoid this increase? Some people would say yes, and they would evacuate even if not required to do so. Some people would stay. The additional risk is small, they would decide, and the loss in giving up their homes is huge.

If you were president, would you feel right forcing people to evacuate, or would you let them decide? The risk seems small, but among a population of 30,000 people, the additional 1.8% would yield 500 excess cancers.

I don't give an answer to this conundrum, because I don't have one. This is not a physics question. Physics can express the consequences of the options, but wise leaders have to make the hard decisions. Is it right to force people to avoid a risk that they consider worth taking? Is it right to allow 500 deaths that you could avoid? How do you reconcile these conflicting issues?

The number of deaths expected is considerably higher when you take into account the distant regions of the world that received small doses of radiation. If 25,000 people got 0.1 rem each, that still adds up to one more cancer death, and the numbers are much worse than that. The map in Figure 8.2 shows an early estimate of the doses received by people in various regions of Europe.

Note that the darkest region is the area where the population was exposed to 1 rem or more. What is the total number of excess cancers expected from the Chernobyl accident? To calculate that value, we must take the number of people living in each dose region, and add up the rem exposure for all those people. For each 2500 rem, we assume (from the linear effect) that there will be one cancer. Because this number is of so much concern to the world community, a great deal of effort was made to measure the distribution of the radiation. In 2006, the International Atomic Energy Agency (the IAEA, associated with the United Nations) came up with its best estimate for the total dose: about 10 million rem. That implies that the total number of induced cancer deaths from the Chernobyl accident would be 10,000,000 divided by 2500, for a total 4000 excess

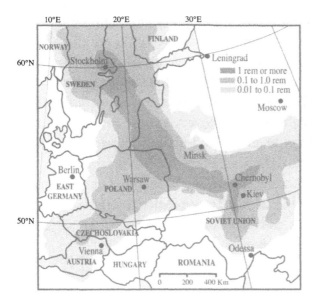

Figure 8.2. Levels of radiation from the Chernobyl disaster that reached various parts of Europe.

cancers. As I said earlier, that's far greater than the 500 expected in the immediate region.

A surprising fact for many people is that even if this calculation is correct, and the number of cancers predicted is accurate, it will be difficult to identify the people who are being killed by the Chernobyl accident. Four thousand additional deaths are hard to detect among a population of millions because cancer from other causes is so common. Certain kinds of rare cancer can be attributed to the accident. Much of the released radioactivity came in the form of radioactive iodine, which concentrates in the thyroid gland and causes thyroid cancer. Thyroid cancer is sufficiently rare that the dozen or so cases observed in the Chernobyl region are almost certainly attributable to the accident. Thyroid cancer is treatable, yet nine of the Chernobyl thyroid cancer victims died. Beyond these, it is virtually impossible to point to any cancer victim and attribute the illness to the release of radiation from Chernobyl, just as with the cancers in survivors from Hiroshima and Nagasaki.

Should we be concerned about deaths that can't be observed statistically? Of course we should. Those 4000 people are all individuals who otherwise would not die from cancer. There is a strange paradox here: tragedy is occurring, and yet it is almost invisible amid the much larger tragedy of cancer from other unknown causes.

Several diseases in the Chernobyl region are worse than this radiation-induced cancer in the sense that they are responsible for many more deaths. Two of these are cancer and heart disease from intense smoking and alcoholism. If evacuation of the region causes stress in the population, and that leads to an increase in substance abuse, that could become the bigger health effect from the accident.

The number of additional cancer deaths may end up being much lower, only 500 instead of the 4000 now predicted. The reason is one of the most important and hotly debated issues in the subject of radioactivity: the linear hypothesis.

The Linear Hypothesis

I already discussed the linear effect. Nobody really disagrees with it. But there is another important issue, which usually goes by the name of the *linear hypothesis*. This is the simple assumption that the straight line we drew in the cancer graph (Figure 8.1) is accurate at all doses, even very small ones. That seems like a reasonable assumption, but it turns out to have huge policy consequences, and it may be wrong. The assumption affects the answer to many questions, including these: How dangerous is nuclear waste? How many people would be killed by a nuclear dirty bomb? What are the long-term consequences of radioactive leakage? At what level of radiation must we consider evacuating a contaminated area?

The linear hypothesis is called a *hypothesis* because we don't yet know if it is true. Let's go back to the cancer dose plot. I'll redraw it, focusing on the low-level part, but this time I'll draw the line a little differently (see Figure 8.3) to illustrate the controversy. Notice that the new line has a kink in it. It shows what the plot would look like if there were a threshold, similar to the one we know about for radiation illness and viral disease. As drawn, the line from 0 to 6 rem

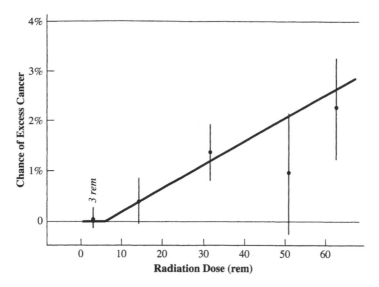

Figure 8.3. Low-dose region of the cancer dose plot, enlarged from Figure 8.2. The bent line is drawn to show what we would expect if there were a cancer threshold of 6 rem.

shows no cancer. Then above the 6-rem threshold, the curve rises linearly—that is, follows a straight line.

Could this bent line be correct, and the previous unbent line wrong? Yes; in fact it does a better job of passing through the data points (and the short vertical lines that represent the statistical uncertainties) than does the previous, unbent line. Some scientists argue that it makes theoretical sense too. Most poisons trigger a *biological defense mechanism* that repairs the damage. In fact, some low doses stimulate and exercise the defense mechanism, making you more resistant to further attack. That which does not kill you can make you stronger.

Does that mean that the bent line is correct and the old straight line was wrong? No, not necessarily. I just said it *could* be wrong, not that it is. Statistically we can't distinguish between the two possibilities, because the error uncertainties are too great. And some scientists offer a theoretical argument for the unbent line. They say that no matter how small the dose, radiation has some chance of causing a mutation, so the number of cancers will be proportional to dose, with no threshold. Cells are being traumatized all the time. That's what gives the 20% rate. Increasing this rate just a little bit will increase the cancer.

What is the truth? Is the linear hypothesis correct or not? And does it matter? If the effect is so subtle, then is it really important to know the answer?

Perhaps surprisingly, the answer is yes. Let's think about Chernobyl again. Using the linear hypothesis, the United Nations estimated the number of deaths around the world caused by radiation that leaked from Chernobyl to be 4000. If there is a 6-rem threshold effect and the linear hypothesis is wrong, then anyone who got less than 6 rem of radiation has a zero chance of excess cancer. Because many people did get higher doses, the total number of cancer deaths is still high—about 500—but much lower than the 4000 estimated from the linear hypothesis. It was tragic, but not nearly as tragic as many other disasters around the world. In fact, maybe the evacuation of the wide region where people would get only 6 rem or less was completely unnecessary. If there is a threshold, then the risk of excess cancer from those low exposures is zero.

The issue is important even in the United States. If there is a threshold effect, then many of the worries about nuclear power vanish. Nuclear waste storage is not a problem, because even if the waste leaks, individuals are unlikely to get a dose of more than a few rem of radiation.

Right now, the US government sets a limit for the general public's exposure to radioactivity. What limit do you think is allowed? Take a guess before you read on.

The answer is 0.1 rem, far below the hypothesized threshold. Moreover, if there is a radioactive spill, the government requires that it be cleaned up until the residual radioactivity gives a dose per year of no more than 0.025 rem. The logic for this number goes something like this: If the linear hypothesis is true, then even 0.1 rem causes cancer, with a probability of 0.11/2500—that is, 0.004%.

If we were to expose the entire US population to that level, all 300 million people, there would be 12,000 additional cancers. That would be very bad. So we set the limit low. Why not? It's easy to be cautious.

The low limit, however, can have serious consequences. Suppose, for example, that a terrorist releases radioactivity over a large area in a major city. Let's assume that the level is 0.1 rem. Then for anyone working or living there, the chance of cancer increases from, say, 20% to 20.004%. Should the government evacuate the area? Or should the government change its previously determined standards? How will the public react to either decision?

All this is made more complicated by the fact that the environment contains natural radioactivity. I'll come back to this point later, but for now consider the fact that someone living in Denver is exposed to about 0.1 rem per year of natural radioactivity more than someone in New York City. Should we evacuate Denver and clean it up? We can't clean it up, it turns out, because the radiation comes from natural radon that leaks from natural uranium in the rock and soil. Nobody seems concerned, because it turns out that cancer deaths in Denver are lower than in other parts of the country, despite the radiation.

By the time you become president, it is unlikely that the validity of the linear hypothesis will have been determined. Even if you were to experiment with a million mice, the effects at 1 rem of radiation are still too small to be observed.[9] The best hope comes from experiments that could elucidate the mechanisms of cancer. The government asked the National Academy of Sciences to review the question, and they published a report in 2006. They examined all the papers that argued for a threshold effect and concluded that the evidence was not sufficiently compelling to change the policy. So the US government continues to derive its laws according to the linear hypothesis.

Future presidents should note that that was a policy decision, not a scientific one. Many people have argued that the effects of the policy are not just in health, but in other central aspects of national policy. Much of the fear of nuclear power, for example, comes from the projections of the number of people who will die of cancer from low levels of radiation. If the policy results in consequences that hurt people in other ways (everything from evacuating their homes to fighting in a war), then the linear hypothesis is no longer the conservative choice. This is policy, not science—at least not yet.

9

RADIOACTIVE DECAY

RADIOACTIVITY does go away, although it can take a long time. Some dangerous radioactivity, such as that of iodine-131, lasts only a few weeks. Plutonium is notorious for lasting 24,000 years, but potassium-40, a key radioactive atom found in food and meat (and especially in bananas), lasts a billion years! Which lifetime is worse? Should we fear more the short-lived or the long-lived atoms?

The issue is complicated, but not too complicated for anyone who aspires to be president. A nucleus can explode only once, and then it is gone. That means that anything containing radioactivity will eventually lose it. As the atoms are used up, the number of remaining nuclei that can explode is reduced. The radioactivity goes away. It *decays* with time. That's why people often refer to the radioactive explosion as *radioactive decay*.

The half-life of a radioactive material is the time it takes for the radioactivity to decay to half of its original level. The following list gives the half-lives of some atoms that come into policy questions.

Polonium-215: 0.0018 second   Strontium-90: 30 years
Polonium-216: 0.16 second   Cesium-137: 30 years

Bismuth-212: 1 hour

Sodium-24: 15 hours

Iodine-131: 8 days

Phosphorus-32: 2 weeks

Iron-59: 1.5 months

Polonium-210: 3 months

Cobalt-60: 5 years

Tritium (hydrogen-3): 12 years

Radium-226: 1620 years

Carbon-14: 5730 years

Plutonium-239: 24,000 years

Chlorine-36: 400,000 years

Uranium-235: 710 million years

Potassium-40: 1.3 billion years

Uranium-238: 4.5 billion years

Note that I said earlier that the radioactivity of iodine-131 lasts a few weeks, yet here the half-life is listed as 8 days. Why the discrepancy? Actually, both statements are correct. This is important to understand because iodine-131 is a major threat and was the cause of all of the identifiable cancers from the Chernobyl nuclear accident. Here's how to reconcile the two numbers. After 8 days, one half-life, half of the radioactivity is gone. You might think that after two half-lives, all of it would be gone, but that's not right. Radioactivity is a probabilistic phenomenon, and the remaining nuclei, those that haven't yet decayed, are the same as they were at the beginning. Even though they are starting their second half-life, they have only a 50% chance of decaying during that second half-life.

A biologist would state it like this: the mortality rate of the surviving atoms does not change with time. That is in striking contrast with human mortality. A person who survives to age 80 has a much higher mortality rate than someone who is 20.

After the second half-life, 25% of the original atoms will be left. In another half-life, half of these will go, and the number will be down to 12.5%. Because only 12.5% of the radioactive atoms are left, the radioactivity is also reduced to 12.5% of its original level. Then 6.25%, and so on. After 10 half-lives (80 days), the amount has been reduced by about a factor of 1000, since that is ½ multiplied by itself 10 times.[10] After an additional 10 half-lives (20 half-lives altogether), the radioactivity is reduced by another factor of 1000, making for a total reduction of a million. Every 10 additional half-lives reduces the radioactivity by another factor of 1000. That fact proves important when we worry about fallout from nuclear bombs, about terror-ist dirty bombs, and about nuclear waste storage.

The key point here is that the danger from radiation drops to ½ after a half-life, but it lingers for many half-lives. The radioactivity is never completely gone until the last atom explodes. But after it is reduced by a factor of a billion (after 30 half-lives), most radiation is completely undetectable and usually harmless.

Let's return to iodine for a moment. Part of the reason it is so dan-gerous is that it decays so rapidly that it gives its victim a large dose in a short time. The iodine concentrates in the thyroid gland, where its radiation triggers thyroid cancer. If you fear imminent exposure to radioactive iodine, you can protect yourself by taking iodine pills (the nonradioactive kind). Your thyroid saturates, and once it has a full supply of harmless iodine, it will not absorb any more. So pop a pill and prevent the radioactive iodine from being absorbed into your thyroid. You need to do this for only a few weeks, because after that most of the radioactive iodine will be gone, dead from decay.

Some people mistakenly think that iodine pills will also protect against waste from a nuclear reactor. But they don't help, because the radioactivity in such waste does not come from iodine. If the waste is more than a few months old, all the radioactive iodine has decayed. The danger from such waste comes from atoms with longer half-lives. Let's get back to our original question. What half-life poses the greatest risk, long or short? It turns out the answer is nei-ther of those.

Dangers of Short Versus Long Half-Lives

In a nuclear accident, the most dangerous materials are often not those with short or long half-lives, but those with intermediate half-lives. That's because atoms with short half-lives disappear rapidly, whereas those with long half-lives take so long to decay that they don't have many decays per second.[11] In fallout from a nuclear bomb, for example, the material that is likely to cause the most human death is strontium-90, which has an intermediate half-life of 30 years. What makes it dangerous is that the half-life is short enough (30 years instead of 1000 years) that it emits most of its radiation in a human life span, yet long enough (30 years instead of 8 days) that we can't easily wait for it all to go away.

Polonium-210, the material used to assassinate Alexander Litvinenko, has a half-life of 100 days. Ponder that number. Can you see why the assassin found it attractive? He wanted to buy the radioactive poison and have it still be potent when he delivered it. If it had a half-life of a week, he might have felt overly rushed. Yet he didn't want too long a half-life, because then the victim might not get a big enough dose to die quickly. So the assassin might have considered a half-life of 100 days to be optimal.

Plutonium-239 from nuclear reactors has a half-life of 24,000 years. This long half-life makes it less dangerous than strontium-90, for the same amount of material. It is a problem for nuclear waste storage, though, particularly for nuclear reactors, because it takes such a long time to decay.

Look again at the list of half-lives. Tritium, the radioactive form of hydrogen that is used to make my wristwatch glow in the dark, has a half-life of 12 years. That means that in 12 years my watch will be only half as bright. After 24 years, I'll probably want to replace it.[12] Tritium is well matched to the life of my watch, just as polonium-210 is well matched to the needs of an assassin.

Now consider the long half-life of uranium-238: 4.5 billion years. We believe that all uranium was created in a supernova explosion that led to the formation of the Earth 4.5 billion years ago. Only half of the uranium is left because it was created one half-life ago. Uranium-235, the kind used in nuclear bombs, has a shorter half-life: about 0.7 billion years. If it was created at the same time, it has now endured over six half-lives, and it should be depleted by a factor of ½ multiplied by itself six times. That's a factor of 64. In fact, the existing amount of uranium-235 is a bit lower than that, suggesting that it wasn't as abundant as uranium-238 when it was created.

Is uranium-238 more dangerous than plutonium because it has a longer half-life? No—just the opposite. In the chapter on nuclear waste we'll talk more about this and, in turn, about people who try to raise concerns regarding plutonium by citing *its* long half-life.

One of the biggest sources of radioactivity in the human body is carbon-14, sometimes called *radiocarbon*.[13] We get our carbon-14 from the food we eat.[14] It has a half-life of 5730 years. In a typical human body, 120,000 nuclear decays occur every minute from carbon-14.[15] Each decay emits a beta ray (a high-energy electron) that damages the surrounding cells. We all live with this internal radioactivity; it is part of every living thing. It is natural, organic, but no less dangerous than artificial radioactivity.

When we die and no longer eat food, the carbon-14 decays and is not replaced. If an archaeologist finds a fossil and measures that the radioactivity is half of that of a living creature, then he knows that the creature died one half-life ago—that is, 5730 years ago. This method is the primary means for measuring ages in archaeology. If the radioactivity is one-fourth the natural level, then the fossil is 2 half-lives old. If it is down by a factor of 1000, then it has been dead for 10 half-lives. Such low rates are very difficult to measure, so carbon-14 is useful only to ages of about 10 half-lives—about 57,300 years, unless advanced methods are used.[16]

Let's return now to an example that appeared at the beginning of the previous chapter. In the United States, alcohol for consumption must be made from fruits, grain, or other plants. Making it from petroleum is illegal. Any alcohol made by the fermentation of plant matter contains recent, radioactive carbon-14. In contrast, petroleum was created by decaying vegetable matter that was buried 300 million years ago. That means that the petroleum was formed from living matter that died more than 50,000 carbon-14 half-lives ago. No detectable carbon-14 remains in it. This absence provides an easy way for the US government to test whether alcohol has been produced from petroleum. The US Bureau of Alcohol, Tobacco, Firearms and Explosives tests alcoholic beverages for carbon-14. If the expected level of radioactivity is present, then the beverage passes the test. If the alcohol is not radioactive, then the testers know that it was not made from grain or fruit, and it is deemed unfit for human consumption.[17]

For the same reason, biofuels—made from corn, sugarcane, or other farmed crops—are also radioactive, whereas fossil fuels are not. The natural radioactivity of the carbon-14 in fossil fuels has long since decayed away. Radiation from biofuels is too weak for it to offer any danger, but it does provide an easy test to see if the fuel is truly made from crops.

Environmental Radioactivity

Presidents sign laws that determine what levels of radiation are acceptable for the population. To do this in a reasonable way, you need to be aware of the natural radioactivity to which we are normally exposed. Not only are you radioactive, but so is the world around you. We already mentioned the radioactivity in you, in your food, and in alcohol, but there is much more. Most of the radiation that we encounter comes from natural sources: potassium in rocks and soil, radiocarbon in the air, natural uranium and thorium. Radioactive radon gas seeps up from within the Earth. Moreover, we get significant radiation from space—cosmic rays coming from distant exploding stars called *supernovas*. A human is exposed to typically 0.2 rem of such radiation each year. That's a much bigger dose than you get from your own body.

This natural dose naturally causes some cancer. After 50 years, your exposure to 0.2 rem per year adds up to 10 rem. Using the linear hypothesis, the expected rate of cancer from this radiation is $10/2500 = 0.004 = 0.4\%$. That's small compared to the 20% observed rate, so we conclude that natural cancer does not come from natural radioactivity.

What is a safe level of radioactivity? If the linear hypothesis is right, then even low doses increase the risk of cancer. If it is wrong, then levels below 6 rem could be completely safe. Earlier I mentioned the natural radioactivity found in the Denver region. Let's look at those numbers again. A reasonable estimate is that the average yearly excess in Denver (compared to the US average) is about 0.1 rem per person per year.[18] For 2.4 million people living in Denver for 50 years, this excess amounts to 0.1 × 2,400,000 × 50 = 12 million rem, enough to cause 4800 excess cancers. That's more excess death than is expected from the Chernobyl nuclear accident!

Now let's see how these numbers affect our hypothetical dirty-bomb incident: A terrorist explodes a dirty bomb in Denver that releases enough radiation to cause 1000 excess cancers. Would you evacuate Denver and save 1000 lives? Remember, the people in this city have already been receiving a far larger dose than most of the rest of the population just by living in that city. Even if there were no bomb, you could save 4800 lives by evacuating Denver right now. Why should 1000 terrorist-caused cancers be treated differently? There is no difference between "natural" radioactivity and terrorist-caused radioactivity. The only thing that counts is the number of rem. If you ever have to make such a decision, you will want to know the numbers. No matter what you decide you will be criticized, so you need to make your decision carefully and be able to justify it.

Some people think that the government should allow the public to be exposed to larger doses than are currently legally permitted, maybe as much as an additional 0.2 rem per year per person, comparable to the natural level. That can't be bad, right? But for 300 million people (the US population), that is a total dose of 60 million rem per year, and if the linear hypothesis is right, that would lead to 60,000,000/25,000 = 2400 extra cancers per year. That's why other people argue that the allowed dose from human-caused radioactivity should be even lower than the dose we get from nature.

Those are the numbers. It will be up to you to decide the policy.

Horrific Mutations

Radiation causes illness and cancer, but it can do other bad things too. Early experiments revealed that intense levels of radiation could cause ghastly mutations in insects. This discovery led to a spate of horror movies. Particularly vivid in my memory were the truck-sized ants in *Them!* (1954; see Figure 9.1), and the giant reptiles in *The Beast From 20,000 Fathoms* (1953) and *Godzilla* (1954). In the original comic, Spider-Man mutated when he was bitten by a radioactive spider, although the more recent movie incarnation substituted a genetically engineered spider. Are such mutations a real danger?

Figure 9.1. Giant ants in the 1954 movie Them!

For higher animals (mammals, lizards, fishes, etc.), no such mutations have been observed. The children of victims in Hiroshima and Nagasaki showed no significant increase in birth defects, according to a 2006 National Academy of Sciences report. That's an important fact for a future president to know, since mutation is so much in the public imagination.

The fact that we don't see the horror-film mutations in higher animals doesn't mean that harmful mutations don't happen. The fetus is particularly susceptible. The United Nations Scientific Committee on the Effects of Atomic Radiation (UNSCEAR) studied the danger.[19] The committee concluded that there is a risk to the fetus of about 3% for each rem of exposure; that's 75 times greater than the risk to adults from 1 rem. That's why pregnant women are well advised to avoid exposure. Mutation in one of the stem cells (cells that can turn into other cells) can lead to mental retardation, malformed growth, or cancer, but usually the result is a spontaneous abortion.

A moving documentary titled *Chernobyl Heart*, released in 2003, attributes a large number of birth defects in the Chernobyl region to the radioactivity from that accident. Most experts think that the film does not accurately represent the truth of the situation, because other similar exposures (such as those in Hiroshima and Nagasaki) caused no similar problems. It is plausible, for example, that the illnesses shown in the movie had always been widespread in the region, perhaps because of the high alcohol and cigarette consumption in that area, but were reported to the outside world only when the region received intense medical attention after the Chernobyl accident. Because the movie was not a scientific study, the makers were under no obligation to present these alternative explanations.

X-Rays and Microwaves

X-rays are the one kind of radiation that everyone in the United States is familiar with. Some people refuse to have dental X-rays because they fear the radiation; ask any dentist. If you agree to the X-rays, the hygienist puts a lead blanket over your body, maneuvers the X-ray machine next to your head, and leaves the room. You hear a short buzz. Then the hygienist returns.

Despite all this drama, tooth X-rays are incredibly safe. The procedures make some patients feel better, but they frighten others. The doses are tiny, less than $\frac{1}{1000}$ of a rem, not much more than your tooth would get by living in the world's natural radioactivity for two days. Should pregnant mothers avoid such X-rays? After all, their fetuses are highly susceptible, so having X-rays taken is obviously not the safest thing to do. But the lead blanket protects the fetus. And the knowledge gained by the X-ray might prevent a serious tooth problem that could, in principle, pose an even greater danger to the unborn baby.

Part of the fear of X-rays comes from historical experience. Doses used to be bigger. As a child in the South Bronx, I had my feet X-rayed in a fluoroscope like the one pictured in Figure 9.2 at my

Figure 9.2. A fluorescent X-ray machine used in shoe stores in the 1950s to help guarantee a good fit.

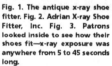

Fig. 1. The antique x-ray shoe fitter. Fig. 2. Adrian X-ray Shoe Fitter, Inc. Fig. 3. Patrons looked inside to see how their shoes fit—x-ray exposure was anywhere from 5 to 45 seconds long.

local shoe store. The dose of radiation that my feet received was probably about 50 rem.[20] That's 50,000 times more than I get from a tooth X-ray today. If the linear hypothesis is correct, then my chance of cancer was increased from a nominal 20% to 20.1% just by one visit to the shoe store.[21] Scientists warned about the cancer danger at the time, but many public officials were slow to shut these machines down because the public loved the devices so much. I even hesitate to say that my exposure in such a store was a mistake. I found the experience fascinating. It is one of the most vivid memories of my youth. Maybe seeing the magic of my toe bones wiggling, and clearly seeing the nails that held my shoes together, helped inspire me to become a physicist.

An X-ray is actually an energetic photon, an energetic particle of light that carries 25,000 times more energy than does ordinary visible light. X-rays pass through water and carbon as easily as visible light passes through glass, but they are quickly stopped by heavier elements such as calcium and lead. An *X-ray photograph* is actually a shadow of the tooth or bone projected on a piece of film. Places with less calcium (for example, decay in the tooth, a break in the bone) allow more X-rays to come through, and that darkens the film. A thick layer of lead in the dentist blanket protects your body (just as lead prevents Superman's X-ray vision from penetrating).[22]

Many people also worry about microwave radiation from cell phones. Unlike X-rays, which are high-energy photons, microwaves are photons with extremely low energy. They deposit their energy in the form of heat; that's what they do in microwave ovens. They do not break DNA molecules in the body (unless they actually burn and char the material), and therefore they pose no risk of causing cancer in the way that X-rays and other energetic radiation (even sunlight) can. The main danger is the heat. Much of the fear of microwaves undoubtedly comes from the fact that they share the name *radiation* with the other, far more dangerous forms, such as gamma radiation. The fear that some people have shown toward such cell phone radiation finds its origin not in physics, but in linguistics.

Is Radioactivity Contagious?

If you are exposed to something that is radioactive, do you become more radioactive yourself? Do you "catch" it, like you catch a cold? In the world of science fiction the answer is yes. In movies, people exposed to atomic bombs come away glowing in the dark. In the real world, the answer is no, it is not contagious—at least most of the time, for most kinds of radioactivity. Remember, radioactivity is the explosion of the nucleus. Such explosions damage your DNA; they usually do not turn other atoms into radioactive ones.

You will become radioactive if some radioactive material sticks to you. Of course, you don't really become radioactive; you just become dirty with radioactive dirt. This could happen if fallout from a bomb landed on you, or if you touched radioactive dust while taking a tour of the inner parts of a nuclear reactor. (When I made such a visit recently, I had to wear special clothing and I was instructed to touch nothing.)

There is an exception—a kind of radiation that can make you radioactive: neutrons. Some forms of radioactive explosions emit neutrons, and when these hit your body they can attach themselves to the nuclei of your atoms and make them radioactive. In reality, to do this would require so many neutrons that you would be dead from radiation illness. Objects exposed to intense neutrons do become radioactive. That happens in nuclear reactors and in nuclear bombs.

The neutron bomb, described in the next chapter, was actually designed to minimize the amount of a bombed city made radioactive. The induced radioactivity it caused was actually very small compared to the typical radioactive dirt spread by more primitive bombs.

Food is sometimes treated with radiation to kill bacteria, viruses, or insects. The food does not become radioactive when this is done. If the radiation is very intense, however, it does damage some of the molecules in the food, so it sometimes alters the taste. Intense radiation could, if overdone, also create compounds that are unhealthy. The same is true for all kinds of preservatives or insecticides. Sterilization by radiation has been studied for many years and has been deemed safe by the UN World Health Organization, as well as by the US Food and Drug Administration and the US Department of Agriculture. As a future president, you will have to cope not only with the facts of radiation, but also with the fear caused by misconceptions about it.

10

Nuclear Weapons

Nuclear weapons are perhaps the most fearsome invention of the twentieth century. They are so important in our culture that we have almost as many names for them as the Eskimos have for snow: *nukes, nuclear bombs, atom bombs, A-bombs, hydrogen bombs, H-bombs, fission bombs, fusion bombs, thermonuclear bombs, uranium bombs, plutonium bombs, neutron bombs, dirty bombs.* During the Cold War, the terms *nuclear annihilation* and *mutually assured destruction* (MAD) were so commonly used that they became clichés. When schoolchildren in the 1950s were asked, "What is your biggest worry?" they typically responded, "The atom bomb." Even today, people hope that nuclear weapons will have no long-term future, but most people fear that they will. The spread of nuclear weapons—nuclear proliferation—is in the news. Iran and North Korea have moved to the center of our foreign policy concerns in large part because they are developing nukes. But to understand the future of nuclear weapons, it is essential to understand first what they are.

There are three basic kinds of nuclear weapons, named after the elements on which they are based: uranium, plutonium, and hydrogen. These three fundamental kinds have very different problems in their construction. Here is a quick overview.

• *Uranium bombs* are made out of a material that is very hard to obtain: purified uranium-235. The purification process requires extreme high tech, such as calutrons, gas diffusion plants, or super centrifuges. If you can get the material, the bomb itself is relatively easy to make. It's the sort of bomb that gives rise to the legend that it could be designed by a high school student. The design is not the hard part. The bigger problem is obtaining the pure uranium-235. A uranium bomb destroyed the city of Hiroshima.

• *Plutonium bombs* are made from material that is relatively easy to get: plutonium-239, a by-product of nuclear reactors. OK, it isn't that easy to get unless you have access to a reactor, but there are a lot of nuclear reactors around the world. If you can get the discharged reactor waste, and if you know enough about radiochemistry techniques to avoid being radiated, then it is relatively easy to extract the plutonium. For a rogue nation or even a terrorist, it is a lot easier to get plutonium than to get uranium-235. As if to compensate, it is very hard to fashion a nuclear bomb using plutonium once you get the material. To do so with plutonium requires implosion, and that is an extremely difficult technology to master. A plutonium bomb destroyed the city of Nagasaki.

• *Hydrogen bombs* can release over 1000 times the energy of the uranium and plutonium bombs, but they are the trickiest of all to make. To start, they require a uranium or plutonium bomb just to trigger the hydrogen burn, so first you have to master that. Then you have to obtain the rare isotopes of hydrogen and lithium that are used. Finally you have to put it all together in such a way that the first bomb triggers fusion rapidly, before it blows the whole thing apart. It probably won't work anyway on the first trial, so you need an elaborate testing program. And testing is all the more difficult if you want to keep your program secret. It's not easy to hide an explosion that releases the energy of a million tons of TNT.

What all three weapon types have in common is enormous destructive power and a severe radioactive legacy. Uranium and plutonium bombs are based on the fact that a large fraction of their atoms can be made to explode in a very short period, typically a millionth of a second. Recall that radioactivity is the explosion of a nucleus, the small inner core of the atom. For uranium-235 and plutonium-239, when the explosion occurs the pieces fly out with about 20 million times as much energy as when a TNT molecule explodes. Since an atom of uranium weighs about the same as a molecule of TNT, that implies that 1 pound of uranium has the energy of 20 million pounds of TNT.[23] According to Robert Serber, one of the key scientists on the Manhattan Project, the Hiroshima bomb used about 80 pounds of uranium.[24] If it had all exploded, the energy released would have been that of 750,000 tons of TNT, nearly a megaton. In fact, however, the energy released was only 13 kilotons.[25] That's because most of the uranium didn't explode. The Hiroshima bomb blew itself apart before all the uranium fissioned. The bomb was only about 2% efficient—pretty much what the Manhattan Project scientists had expected.

The Key to an Explosion:
The Nuclear Chain Reaction

Nuclear weapons require a special kind of uranium called *uranium-235*, or a special kind of plutonium called *plutonium-239*.[26] Ordinary uranium does not work; neither do many kinds of plutonium. When the nucleus explodes, the huge energy comes from the enormous forces that hold the nucleus together. When a nucleus is broken, the pieces fly apart with great energy. Most of this energy comes from the electric repulsion of the fragments. To split the nucleus of uranium, all you have to do is hit it gently with a neutron. The energetic breakup is called *fission*, named in analogy to the fission of a biological cell. Typically, two large fission fragments and two additional neutrons fly out. The kinetic energy of the fission fragments is turned into heat, vaporizing all nearby material, and that's what makes the explosion.

When the nucleus fissions, the extra neutrons that come out can trigger fissions in other nearby nuclei. Four more neutrons emerge, followed by 8, 16, 32, and so on. Each doubling is called a *generation*. The whole process is called a *chain reaction*. The chain reaction can happen so fast that in a millionth of a second, a very large fraction of all the atoms fission.

Let's calculate doublings for 81 generations. That's not as hard to do as it sounds, if you use either a spreadsheet or a pocket calculator. Try it. Start with the number 2, and multiply it by 2 over and over. After 81 doublings, the number of neutrons is 2×10^{24}. That's 2 followed by 24 zeros. That's also the same as the number of uranium nuclei split in the Hiroshima bomb. Think about that. Only 81 doublings, taking less than a total of a millionth of a second, split an enormous number of nuclei and released all that energy.

If you like math, think about this: every new generation will release as much energy as all the previous generations put together. For example, the sum $1 + 2 + 4 + 8 + 16 + 32$ adds to 63. That's just one less than 64, the next number that would be on the doubling list.

That rule has interesting consequences for bomb design. I said that the Hiroshima bomb blew itself apart when only 2% of the fission had completed. It released 13 kilotons of energy. Suppose that the chain reaction had continued for just one more generation. Then the energy released would have doubled, from 13 to 26 kilotons of TNT. No wonder it's so difficult to predict the yield of a bomb.

Although nuclear explosions release radioactivity, most of the death and damage from the Hiroshima and Nagasaki bombs were not caused by radiation, but by the enormous energy released. That energy created a huge fireball of high-pressure hot gas that rapidly expanded, resulting in an explosion and shock wave that destroyed buildings and other structures. Eventually the two cities were turned to ashes. That result was not from the initial explosion, but from the firestorm ignited by the enormous heat. The hot fireball rose up into the atmosphere and produced the famous mushroom cloud that became a symbol of nuclear explosions. Somewhere between 35,000 and 150,000 people were killed in each city. (Since this is a physics book, I don't want to get involved in the ongoing argument about the exact number of deaths. Either value is horrific.) The cloud from the Nagasaki explosion is shown in Figure 10.1.

The mushroom shape actually has nothing to do with the nuclear nature of the event. Any large explosion will give a similar cloud. In Figure 10.1, the horizontal spreading halfway up the photo comes

Figure 10.1. Mushroom cloud from the explosion of the plutonium bomb over Nagasaki.

from the passage of the cloud through the tropopause, the region where thunderheads usually spread out too. The cloud was so hot that it continued to rise even into the stratosphere. The perspective of the photo shows that it was taken from an airplane.

World War II and the Manhattan Project

The Hiroshima and Nagasaki bombs were created during a frantic period in the early 1940s. The United States was in the midst of World War II and knew that such a bomb might be possible. The enemies at that time, Germany and Japan, had some of the best nuclear scientists in the world, and the United States feared that they might be developing such bombs too. A huge and expensive program was set up in the United States and dubbed the *Manhattan Project*. It had nothing to do with Manhattan; that was just a code name. The bulk of the work was done in Los Alamos, New Mexico. You can read all about it in an excellent book called *The Making of the Atomic Bomb* by Richard Rhodes. The scientific leaders of the Manhattan Project included Robert Oppenheimer and Ernest Lawrence. The military leader, Leslie Groves, also deserves a great deal of credit for allowing the scientists to proceed using the same open scientific approach that they used in their normal research, and for giving them just about everything they asked for. The scientists realized that one require-ment for keeping a chain reaction going was to assemble enough ura-nium—a critical mass—so that every neutron emitted in a fission would be likely to hit another nucleus and trigger another fission before it leaked out.

Critical Mass—The First Atomic Secret

Neutrons pass through matter as easily as light passes through glass. Most of the space in the atom is taken up by electrons, and they are basically invisible to neutrons. That's because neutrons don't have electric charge, and electrons don't have nuclear charge (the origin of the nuclear force). Only when a neutron hits a nucleus, with its strong nuclear force, is it stopped. The nucleus is a tiny part of the atom, so most neutrons sail right through most atoms. The chain reaction will not continue if many neutrons leak out without hitting a nucleus. The easiest way to guarantee a hit is to make the uranium thicker. The problem is similar to that of hitting a tree with a rifle bullet. Aim at one tree, and you'll likely miss. Aim into a forest, and you're pretty certain to hit some tree.

The required amount of uranium to guarantee a nuclear hit is called the *critical mass*.[27] This number is extremely important for a real bomb project, because uranium-235 is very hard to obtain. If the critical mass is large, you can't get enough uranium or plutonium to make a weapon. In the Manhattan Project, the scientists initially estimated that the amount needed for a critical mass was about 440 pounds.[28] The Manhattan Project scientists knew they couldn't get that much during the war. They realized, however, that they could surround the uranium with material that they called a *tamper* that would reflect some of the neutrons. This simple idea was apparently missed by the German bomb project scientists, and it may be the reason why they abandoned a bomb effort. With a tamper, instead of leaking out, the neutrons are reflected back in, so the critical mass needed for an explosion dropped by about a factor of four, down to only 33 pounds. That was far more feasible. In fact, they used more uranium than that in the actual bomb to get a bigger explosion.

Plutonium bombs have a smaller critical mass, about 13 pounds, partly because plutonium's fission emits three neutrons (versus two for uranium), so the chain reaction spreads faster. How many generations does it take to fission 2×10^{25} nuclei? This time, multiply 3 by itself until you get this number. The answer is only 51 generations, rather than the 81 required for uranium.

The first atomic bomb, tested at the Alamogordo site in New Mexico on July 16, 1945, used 13 pounds of plutonium-239. Plutonium is so dense that this much would fit in a container that holds

12 fluid ounces, the size of a soft-drink can. The critical mass for uranium, 33 pounds, will fit inside a quart milk container.

It is an interesting historical note that the uranium bomb design, the one eventually used on Hiroshima, had never been tested prior to its use in war. It was the plutonium bomb design, not the uranium one, that was tested at Alamogordo. The uranium bomb wasn't tested for two reasons. The first was that the difficulty with uranium is purifying U-235. We'll discuss methods to do that shortly. By mid 1945, the scientists had separated enough uranium for just one bomb. If it had been exploded in a test, there would have been no uranium left for a second one. The second reason is that the design of a uranium-235 bomb is relatively simple. The scientists of Los Alamos were quite confident that the bomb would work on its first try.

Most of the numbers that I'm citing come from a book by Robert Serber called *The Los Alamos Primer*. Serber was a physicist at the Manhattan Project, and he prepared a series of lectures to teach arriving physicists what the Manhattan Project scientists had figured out about bomb design. These lectures were later assembled into this primer, and it has been republished by the University of California Press. It shows what the physicists were thinking at the time. Some people think harm is done by publishing this book, since it shows clearly the first steps that are necessary for making an atomic bomb. Others argue, however, that the book probably discourages many groups, maybe even countries, from even trying, when they see the difficulty of the project and the sophistication needed to make a good design. Serber's book puts forth the general principles, but the detailed engineering methods are not discussed.

Hiroshima—The Simple Bomb Design

All nuclear weapons work on the same simple principle: First carry fissionable material, uranium-235 or plutonium-239, in pieces spread far apart from each other. With such separation the neutrons from a fission are likely to escape through the gaps, so the chain reaction doesn't grow. Then, when the bomb is over the enemy target, bring the pieces together, a process diffidently called *assembly*. You have to do this rapidly because when the pieces get close, some of the neutrons start hitting nuclei, generating heat—heat that might blow the pieces apart before the entire mass is fully assembled. Miss out on only a few generations, and the explosive might of the bomb is greatly reduced.

The nuclear bomb that destroyed Hiroshima used uranium-235 and a simple "gun"-type mechanism. By *gun*, I mean that a piece of uranium-235 was shot by a cannon at another piece of uranium-235; the combination but not the individual pieces exceeded the critical mass. The cannon was only 6 feet long and weighed less than half a ton. Figure 10.2 shows the Hiroshima bomb. It was small enough to fit in the bomb bay of a B-29 bomber. The cylindrical form is due to the shape of the cannon inside. The energy released from the fission chain reaction was about 13 kilotons of TNT equivalent. The day after Hiroshima was destroyed, President Harry Truman mistakenly

Figure 10.2. The Hiroshima bomb. It was 10.5 feet long, and its cylindrical shape was due to the 6-foot cannon inside that shot two pieces of uranium together to make a critical mass.

announced that the yield was 20 kilotons. It turns out that he had confused the yield of the Hiroshima bomb with the yield of the test explosion at Alamogordo, New Mexico.

The Hard Part: Purifying Uranium-235

The difficulty in making a Hiroshima-style bomb is not in the design or construction. The gun design is very simple, perhaps even within the means of small terrorist organizations. The hard part is obtaining the pure uranium-235. Natural uranium contains a bomb pollutant that is very hard to eliminate. Programs to purify the uranium typically cost billions of dollars and require top scientists working with advanced technologies.

The pollutant is another form of uranium: uranium-238.[29] The problem with uranium-238 is that it absorbs neutrons, sometimes making—it turns out—plutonium, but in the process it doesn't emit additional neutrons. Just as a fire won't burn if you take away the air, a chain reaction won't proceed if you take away the neutrons.

Ordinary uranium mined from the ground contains 99.3% pollutant uranium-238, and only 0.7% of the fissionable uranium-235. In other words, natural uranium is mostly pollutant. Separating U-235 from the U-238 is called *uranium enrichment*. When the enrichment nears 100%, the uranium is called *bomb-grade*, and the removed material is called *depleted uranium*. Depleted uranium is uranium-238 with less than 0.7% uranium-235.

Uranium enrichment is extremely difficult. Because U-235 and U-238 are both forms of uranium, ordinary chemical techniques don't distinguish between them. Enrichment methods must take advantage of the fact that uranium-235 atoms are slightly lighter than those of uranium-238. But the difference is tiny, just 1.3%.

During World War II, the bomb project leaders at Los Alamos tried several different approaches for enriching uranium-238. The method that succeeded first was the *calutron*, invented and built by Nobel Laureate Ernest Lawrence. He named it after Cal, the nickname of the University of California at Berkeley, maybe because it was shaped like the letter C. Calutrons built at Oak Ridge, Tennessee, enriched essentially all of the uranium-235 used in the Hiroshima bomb. Figure 10.3 shows a calutron used during World War II.

Figure 10.3. A calutron, the device used to enrich uranium for the Hiroshima bomb.

A calutron works by vaporizing the uranium and accelerating it in a magnetic field along the C shape seen in the photo. If they are moving at the same speed, the heavier uranium-238 atoms tend to travel in a 1.3% larger circle than do the uranium-235 atoms, and at the end of the path they are separately collected. The method is very slow and tedious, but after a year of operation, the calutron had managed to separate enough uranium-235 for the Hiroshima bomb.

After World War II, a completely different method, called *gas diffusion*, was used to purify uranium. In this approach, uranium is combined with fluorine to make uranium hexafluoride, a compound that becomes a gas at the relatively low temperature of 134°F. This gas is pressurized to make it diffuse through a porous material. The lighter molecules (with uranium-235) tend to move faster, so they diffuse more rapidly. Each stage of purification increases the enrichment only a tiny amount. As a result, uranium diffusion plants tend to be very large to accommodate the many stages required. The plant in Oak Ridge, Tennessee, shown in Figure 10.4, covers a good fraction of a square mile. But it was gaseous diffusion that was used for virtually all of the enrichment that the United States used to build up its nuclear arsenal during the Cold War.

One of the most highly classified secrets of the diffusion enrichment method was the nature of the special porous material used. It had to survive against the very corrosive behavior of the uranium hexafluoride. The identity of this material has now been made public: Teflon! Many people mistakenly think that the first major use of Teflon was during the space program, but in fact Teflon was the key material in the uranium diffusion plants.

Figure 10.4. The uranium diffusion plant in Oak Ridge, Tennessee.

Centrifuges

The most modern and efficient method for separating uranium-235 is to use a gas centrifuge. You know what a centrifuge is like if you've been on amusement park ride in which you stand against the wall of a large cylinder, which is then rapidly spun. You feel pressed against the wall, pushed there by the centrifugal force.[30] Then suddenly, to your surprise, the floor drops away, but you stick to the walls, held up by friction. A gas centrifuge works on the same principle, but instead of pushing people to the walls, it pushes uranium hexafluoride, the same gas used in the diffusion plant. The gas with the heavier uranium-238 tends to concentrate on the outer part of the cylinder, leaving the lighter uranium-235 near the center. The U-235 is then piped out, as shown in Figure 10.5. In fact, the fractional enrichment in one centrifuge is small (because the U-235/U-238 mass difference is small), so the gas must be pumped through thou-

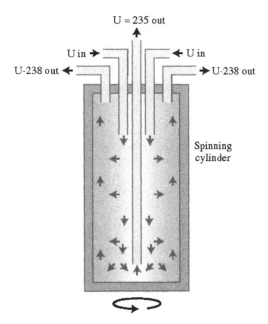

Figure 10.5. Gas centrifuge for separating uranium-235 from uranium-238.

sands of centrifuges in order to get enough enrichment for nuclear power plants or nuclear weapons.

Although you need a lot of them, modern centrifuges can be made efficient and relatively small. Because they spin so fast, they must be made of very strong materials to keep from breaking apart. One of the key new materials, *maraging steel*, is used primarily for uranium centrifuges, rocket bodies, and high-performance golf clubs. US intelligence services become very suspicious of countries that begin importing or making substantial quantities of maraging steel, unless they are major manufacturers of golf equipment.

A typical centrifuge plant has several thousand centrifuges, but the entire collection fits comfortably inside a space no larger than a movie theater. Such systems can produce enough enriched uranium for several nuclear bombs per year. Hidden centrifuge plants are very difficult for intelligence agencies to locate. They don't require large amounts of power, and they are very quiet, thanks to the exquisite balancing of the cylinders that helps keep them from spinning themselves apart.

Gas centrifuges have been the method of choice for nuclear proliferation. Centrifuges were used for enrichment in Pakistan by Abdul Qadeer Khan, and the technology was shared with other developing countries, including North Korea and Libya.

The Plutonium Bomb—An Implosion Design

Even though both uranium-235 bombs and plutonium-239 bombs use nuclear chain reactions, many technical details make the two technologies remarkably different. It is important for future presidents to know and understand these differences because they will help you recognize and evaluate possible proliferation and terrorist threats.

Plutonium is created in most nuclear reactors (see the nuclear power chapter), including those built to produce electric power. It comes out mixed with the other nuclear waste, but it can be separated using relatively straightforward chemistry. When this separation is done, we say that the spent fuel from the nuclear reactor has been *reprocessed*. That's a key word that you'll see in the newspapers. Developing nations that signed the NPT (the Nuclear Non-Proliferation Treaty) have agreed that they will not reprocess spent fuel—that is, extract the plutonium created in their nuclear reactors.

Reprocessing is so simple compared to uranium enrichment, and nuclear reactors are so common, that plutonium bombs seem to be the bomb of choice for small countries such as Pakistan and North Korea. But plutonium bombs are not easy to design. They tend to *predetonate* (that is, explode prematurely) because of a radioactive pollutant that is also created in the reactor. That is probably why the North Korean test of 2006 fizzled. It yielded less than a kiloton of explosion, even though it required a full critical mass of plutonium.

The pollutant in the plutonium bomb is plutonium-240, heavy plutonium. It is the same as plutonium-239, except that it has an extra neutron in the nucleus. That makes it highly radioactive. Plutonium-240 fissions spontaneously; it does not have to be hit by a neutron. This spontaneous fissioning releases neutrons at random times. You might think extra neutrons would help, but in fact they hurt. They cause a partial chain reaction to start even before the plutonium is assembled into a full critical mass, and this partial chain reaction explodes the bomb prematurely and stops the chain reaction before much energy is released.

This problem was recognized during World War II, and for a while the Manhattan Project scientists thought that a plutonium bomb would prove impossible. But a remarkable solution was found: implosion. Instead of bringing the two pieces together with a cannon, take a shell of spread-out plutonium and surround it with explosives. The explosion around the plutonium tends to drive it in—that is, implode it—and compress it into a very tight mass. When the plutonium is so strongly compressed, the atoms are so close together that the neutrons have less space to leak out. If the compression is done properly, this condition is reached so rapidly that the plutonium doesn't have time to predetonate.

It is extremely difficult to make an implosion work. Some people compare the process to trying to compress a water balloon in your hands. Squeeze a little bit, and it tends to squirt out through your fingers. The implosion must be very precise. The high explosives used must be extremely uniform, and manufactured in exquisite balance on each side of the plutonium. They have to be detonated simultaneously and uniformly. That is very tricky, since explosives are usually detonated at specific points, not over extended surfaces. A special technique was developed to take a point explosion and turn it into a uniform implosion. It was called an *explosive lens*. The high explosive would be detonated at a point, but then focused to be smooth over the surface of the plutonium.

Getting all that to work properly is much more difficult than building a suitable gun for a uranium bomb. It requires top scientists and engineers, tricky technologies (such as precision machining of high explosives), lots of testing, and lots of money. For all these reasons, it is unlikely that a plutonium bomb could be built by a small organization such as a terrorist group. The full resources of a country (such as Pakistan or North Korea) are probably necessary. Even then it might fizzle. The North Korean test of 2006 yielded less than a kiloton of explosion, compared to the 20 kilotons of the US Alamogordo/Nagasaki design. Most experts suspect that the low yield on the North Korean test was the result of just such a fizzle, and that predetonation blew the bomb apart before the chain reaction could reach the desired levels.

Figure 10.6 shows the Nagasaki bomb. Notice that the bomb is fatter than the Hiroshima uranium bomb. That shape difference reflects the spherical shell of explosives for the implosion.

Let me summarize some important facts: 33 pounds of uranium-235 is enough for a nuclear weapon. If the chain reaction proceeded by doubling, all the atoms would be split in only 81 generations. Plutonium takes only 13 pounds, no more than it takes to fill a soft-drink can. It can release all its energy in just 51 generations because each fission releases typically three neutrons instead of two. The Hiroshima

Figure 10.6. The Nagasaki plutonium bomb. The round shape is due to the spherical shell of explosives required for implosion.

bomb was a gun design using uranium. The Nagasaki bomb, similar to the one tested in New Mexico, was a plutonium implosion bomb. Future presidents need to know these facts because they are important for evaluating the prospects of nuclear proliferation.

The Hydrogen Bomb

The most destructive bombs of all are hydrogen bombs, also called *H-bombs* or *thermonuclear bombs*. The largest ever exploded released the energy of 55 *million* tons of TNT, over 4000 times more than the Hiroshima bomb. Most of the US nuclear arsenal consists of hydrogen bombs. They can be made so small that 14 can fit on the top of a submarine-launched Poseidon intercontinental ballistic missile. Figure 10.7 shows hydrogen bombs being assembled to fit in a warhead. The cone shape of each warhead is designed to allow them to be accurately aimed at targets even when they reenter from space.

Figure 10.7. Cone-shaped hydrogen bombs being assembled to fit in a missile warhead. Each one could destroy much of a major city.

The assembly is called MIRV—for Multiple Independently-targeted Reentry Vehicles. Independently-targeted means each man-sized warhead can destroy the heart of a different major city.

The hydrogen bomb is a two-stage device. First, an "ordinary" uranium or plutonium atomic bomb is exploded. The radiation from this primary stage heats a specially designed secondary stage containing two special forms of hydrogen: heavy hydrogen, called *deuterium*; and doubly heavy hydrogen, called *tritium*.[31] At very high temperatures, these forms of hydrogen will *fuse* into helium and release great energy—more per pound than uranium or plutonium does during fission. Fusion is the same process that takes place deep in the sun to provide all the power of sunlight.

The hard part, as it was for the plutonium bomb, is to make the process take place very rapidly, before the hydrogen is blown away by the fission primary. Imagine how difficult this is. In the fission bombs, all you had to worry about was the energy of the high explosives smashing the bomb to smithereens; with a fusion bomb you have to get everything done before an atomic bomb blasts it all to vapor.

For many years, scientists thought that such a design might prove impossible. Then a trick was devised—the "secret of the hydrogen bomb." The secret is this: a plutonium fission bomb emits enough X-rays that they can be used, after bouncing off uranium walls, to compress and ignite the hydrogen. The X-rays travel at the speed of light, so they reach the hydrogen before the blast of the bomb blows it apart. This trick was classified until recently; the exact geometries to accomplish this are still classified. As with a plutonium bomb, getting the implosion to be completely uniform is very difficult. Hydrogen bombs are extremely difficult to design.

Because it depends on fusion—that is, the bringing together of two hydrogen nuclei—the hydrogen bomb is sometimes called a *fusion bomb*. Because it depends on high temperature, it is also called a *thermonuclear bomb*. So *H-bomb*, *hydrogen bomb*, *fusion bomb*, and *thermonuclear bomb* are all different names for the same thing.

A second secret has been public for a longer period: Instead of using tritium, the radioactive form of hydrogen as one of the key fuel components, the bomb can contain a stable (not radioactive) form of the metal lithium called *lithium-6*. This is combined with deuterium (heavy hydrogen, which is not radioactive) to make a solid compound called *lithium deuteride*, or *LiD*. This material is stable and dense, making it an ideal fuel. When the plutonium primary detonates, the neutrons from the fission weapon break up the lithium-6 to make tritium. Thus, half of the fuel is created in the same microsecond that the bomb is exploding. The other half, the deuterium, is found in nature and readily extracted from ordinary water.

I mentioned that fusion emits neutrons. In a typical hydrogen bomb, these are used to yield even more energy, by surrounding the secondary with depleted uranium—that is, uranium-238. (This uranium also reflects X-rays.) The neutrons from the fusion have high energy, and uranium-238 tends to fission when hit with such particles. Typically about half of the energy from a hydrogen bomb comes from this added fission. The fission of the uranium-238 also enormously increases the number of fission fragments, and these create the great danger known as *fallout*. Fallout from a hydrogen bomb can easily kill more people than the blast itself.

If uranium-238 can fission, why can't it be used by itself for a fission bomb? The answer is that it does not produce enough neutrons when it fissions to keep the chain reaction going. But its fission still does release energy, and fission fragments.

Fallout

Much of the danger from megaton nuclear weapons comes from the nuclear fallout. Fallout consists of the fission fragments from the uranium and plutonium in the bomb. It is particularly bad if the bomb is exploded near the ground, since then a lot of dirt and other materials are caught up in the fireball of the explosion. This material, which is not itself radioactive, rises in the air along with the highly radioactive bomb debris. If it all stayed high in the air, the radioactivity of the fission fragments would not cause great harm. Much of the initial radioactivity consists of elements with short half-lives, so if these elements stay up for several hours, they decay away. But if a lot of dirt from the ground is mixed in, then its weight tends to make the combination fall, and that brings the radioactivity to the ground. Such fallout can spread over large areas and kill more people than the blast itself does.

In the 1950s, fallout shelters were established all over the United States. These were areas underground that had enough supplies of food and water that people could live there for a month or so while waiting for the short-lived radioactivity of the fallout to expend itself, or at least fall below the threshold for radiation illness. Many people mocked these shelters because it was clear that they could not survive a direct blast—but that was not their purpose. Others imagined the horror of people fighting over access to such shelters, denying entrance to latecomers because of the limited supplies. Some people built their own fallout shelters, complete with weapons to defend themselves from intruders. As if the horrors of nuclear war were not awful enough, these scenarios seemed to exacerbate them.

Although most of the initial radioactivity from fallout comes from short-lived radioactive elements, more than 5% of the fission fragments are made of the isotope strontium-90, a highly radioactive material that has a half-life of 29 years and can readily enter the food supply. Back in the 1950s, when many people were worried about the long-term effects of nuclear testing, the term *strontium-90* was well known to the general public. Sr-90 falls on grass, is eaten by cattle, is passed on in cow's milk to children, and becomes concentrated in the bones because its chemistry is similar to that of calcium.

Another widely mocked measure of the 1950s was the "take cover" drill carried out in schools. As a child, I was trained to be on the alert for a bright flash—a sign of a nuclear bomb explosion—and if I saw one, to "duck and cover," under a desk or table if possible. No wonder children cited atom bombs as their main fear; they were *told* that this was their main fear. Again, people made fun of the drills because a desk could not possibly protect a child from a nuclear bomb, but they misunderstood the purpose. Even many miles from the center of the explosion, the blast wave of air could still shatter windows. A desk gives important protection from flying glass.

Big Bombs

Making large fission bombs proved to be very difficult. They tended to blow themselves apart. Uranium and plutonium were expensive. In contrast, deuterium and lithium-6 were relatively cheap, and the design evolved to the point where a hydrogen bomb could be made with almost arbitrarily large explosive energy. The early 1950s became the era of the "Big Bomb." The United States and the Soviet Union competed to see which country could make the biggest blast. The Soviet Union won that "honor" in 1956, when it exploded a bomb with the energy of 55 million tons of TNT, over 4000 times more energy than had been used to destroy Hiroshima.

The race was abandoned, in part because of popular pressure

building from the fact that the tests were polluting the Earth with radioactivity. In fact, the radiocarbon in the atmosphere had doubled just from these tests. But more cynical observers suggest that the real reason the tests were abandoned was that the bombs were too big to be carried by missiles. And a bomb divided into smaller parts can usually do more damage.

Smaller Weapons, More Destruction

Many people are surprised to learn that more destruction is possible when big bombs are broken up into little bombs. But when you divide the explosion, you get to spread it out more. A big bomb can vaporize a large area, but a bunch of small bombs can totally destroy (but not vaporize) a much larger area.

Here's the way it works. Look at the image in Figure 10.8 showing the destruction that would be caused by a 1-megaton bomb in New York City. The inner circle is the region destroyed by blast; the outer circle shows the region devastated by fire. For a 50-megaton blast, the radius of destruction would be almost four times larger and would cover most of the region visible in Figure 10.8.[32]

Now let's use that fact to see what happens when we divide a bomb into smaller ones. Suppose we take one big bomb with a destructive radius of 2 miles. The area destroyed is $\pi R^2 = 12.6$ square miles. Now divide the same bomb into eight smaller bombs. The yield of each is $\frac{1}{8}$ as big, and the radius of destruction of each will be only $\frac{1}{2}$ as great,[33] about 1 mile each. Each bomb will destroy $\pi R^2 = 3.14$ square miles. There are eight of them, so if you spread them out over the city, you can destroy $8 \times 3.14 = 25$ square miles of the city—twice what you would get from the single bomb.

There is a simple way to think about this otherwise surprising effect. The one big bomb is equivalent to exploding the eight little ones all at the same place at the same time. That is a waste of destructive energy, since even one of the little bombs is enough to destroy everything out to $\frac{1}{2}$ mile. Putting eight bombs at the same

Figure 10.8. Expected destruction caused by a 1-megaton nuclear strike against New York City. The inner circle shows the region devastated by blast, and the outer circle shows the region destroyed by fire.

place can't destroy that ½ mile any more. If your goal is to kill, then spread the destructive energy over a larger area.

In the 1980s, President Reagan announced that he had unilaterally reduced the total yield of the US nuclear arsenal. What he did not say publicly at that time was that, even though the total yield was down, it had been distributed among many smaller weapons, resulting in an increase in the total destructive capability. He was being honest, but not candid. Or maybe he didn't understand the physics as well as you will when you become president.

11

NUCLEAR MADNESS

IN EARLY 2007, the United States and Russia disclosed to each other the number of nuclear strategic warheads that they had operationally deployed. The number for the United States was 5866. Russia said it had 4162. Most of these warheads are less than a megaton, but even so, it seems like madness to have such destructive energy ready to launch. To some people, the situation is analogous to going shopping while wearing dynamite strapped to your chest. If anyone tries to rob you, you could threaten to blow him (and you) to smithereens. That makes you safe, doesn't it?

Yet these levels met the terms of a treaty signed in 1991, in which both countries promised to reduce their stockpile of nuclear weapons to less than 6000 each. The treaty is scheduled to expire in December 2009. In a treaty signed in 2002, each side pledged to reduce the warheads further, to about 2000.[34] For all these weapons, there are about 10 different "designs," but most involve a combination of fusion and fission.

How many is enough? The historical reason for the huge stockpile is interesting. During the Cold War, the United States feared a surprise attack from Russia's predecessor, the Soviet Union (a replay of Pearl Harbor?), and assumed that most of its weapons would be destroyed in such an attack. It wanted to make sure that even if only 1% of its weapons survived, they would be enough to destroy the Soviet Union. Of course, that requires a 100-fold overkill capability. The United States assumed that if the Soviets knew that retaliation was inevitable, they would never attack. The concept was called *mutually assured destruction* or MAD. Many people consider this policy to be truly mad, as its acronym suggested. Others consider it to be a successful policy, the reason why war between two superpowers—the United States and the former Soviet Union—never took place. See the 1964 movie *Dr. Strangelove* for an ironic account of the dangers inherent in such a strategy.

The answer to the question "what is enough?" depends not so much on physics, but on the vision of what the future strategic situation will be. Some physicists argue that the danger of nuclear holocaust is so great that the only moral approach would be to reduce the weapon stockpile to almost zero. Others counter that war always looks terrible, until you really discover that your freedom is at risk. These are not physics questions, so I won't explore them further here.

The big technical issue today is *stockpile stewardship*. This term refers to the fact that as our weapons grow old, some people worry that when called upon to explode, they may fail to do so. The implosion of the fission primary is a very delicately balanced phenomenon, and if the properties of the materials used in the design change with age, the implosion might not go as planned. In the olden days, we assured ourselves that the implosion was still working by periodic testing, but we have now entered an era when we have decided to end all such tests. (This is partly an attempt to keep other nations from developing nuclear weapons.) So there is a large program at Livermore and Los Alamos to develop methods of testing the reliability of the weapons without having to set off any of them. It is a big technical challenge.

The Neutron Bomb and the Weapons Paradox

A variation of the hydrogen bomb that achieved some notoriety under the Carter administration is the *neutron bomb*. I mentioned that hydrogen fusion produces high-energy neutrons. Such neutrons can kill by causing intense radiation illness. The idea of the neutron bomb is to create a bomb that has lots of neutrons, to kill people, but releases relatively little energy and few fission fragments. The neutron bomb was seen as a way to repel Russian troops if they invaded a friendly country. We would want to kill enemy soldiers but not destroy the friendly country.

The trick to making a neutron bomb was to have a very weak primary stage to minimize fission fragment production, and a small secondary so that the blast wave would not do much damage. Neutrons would still travel hundreds of yards and kill troops.

When the idea of the neutron bomb was made public, many people reacted with horror. The problem, ironically, was that the neutron bomb was less terrible than an ordinary hydrogen bomb. You might think that would make it more acceptable, but just the opposite was true. The fact that it did not destroy buildings meant that the neutron bomb was more likely to be used. That "lowered the threshold" for use of nuclear weapons and therefore was considered immoral.

I consider this to be an example of something I call the *weapons paradox*: Invent a weapon that is more terrible, and it is obviously immoral. Invent a weapon that is less terrible, and it, too, is immoral —because it is more likely to be used. The odd conclusion is that you can't propose any change in armaments without being called names. Examples of this paradox are found even outside the military—for instance, in the proposed use of Tasers by the police to stun suspects.

Nuclear Proliferation

The Manhattan Project scientists and engineers built three nuclear weapons in just three years, once they began serious efforts. So it couldn't really be too hard to do, right? Of course, the Manhattan Project included most of the world's top scientists. But now much of the physics they had to invent has been openly published. The fundamentals of uranium and plutonium weapons design are available in *The Los Alamos Primer* (Figure 11.1), a text written by Manhattan

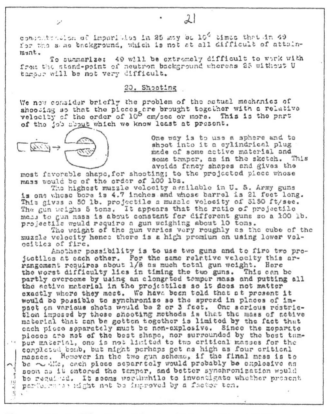

Figure 11.1. The Los Alamos Primer *(page 21), as originally declassified and published by the United States during the 1950s Atoms for Peace program. This page describes the possible gun-style uranium bombs.*

Project physicist Robert Serber. The document was declassified during the Atoms for Peace program under the Eisenhower administration, and circulated as a mimeographed paper. The page in Figure 11.1 was reproduced from a copy that I checked out from the Lawrence Berkeley National Laboratory library. A hardbound edition, with updated comments, has now been published by the University of California Press. It is an interesting book to peruse, both because it could save any serious group a lot of work, and because it illustrates the sophistication necessary to make a bomb work. It shows, for example, how the critical mass necessary for a bomb can be substantially reduced by a series of tricks, including the use of a tamper mentioned earlier.

Why didn't Germany develop a nuclear weapon during World War II? Some experts believe that the Germans failed to find some of the solutions described in the primer, incorrectly thought the critical mass was huge, and for that reason prematurely gave up their attempt to make such a bomb.[35]

A series of other countries have now developed and tested nuclear weapons, including the United Kingdom (which was privy to our nuclear secrets during World War II), Russia, France, China, India, Pakistan, and recently North Korea. A secret nuclear facility in Syria became publicly known when Israel attacked it in September 2007. In 1973, Israel was accused of having conducted a secret nuclear test explosion, and I served on a White House committee to study the intelligence data. Our group concluded that the evidence for the purported test was not valid. Even so, most experts believe that Israel has constructed such a bomb; they note that all three Manhattan Project bombs exploded the first time they were used, so tests aren't absolutely necessary if care is taken in the design. Iran has been developing centrifuge enrichment facilities that could be used to purify uranium-235 for a weapon.

The easiest bomb to build would indeed be a uranium-235 gun-style design. If the uranium could be purified, no complex implosion method would be necessary. Indeed, Saddam Hussein was developing a uranium bomb prior to his invasion of Kuwait. After that war, the United Nations discovered that he had built a series of calutrons to enrich uranium. The best estimate is that he had achieved about a 35% enrichment. The remaining steps to bomb-grade U-235 are easy, and he soon would have had a bomb. However, the United Nations destroyed the Iraqi calutrons, as shown in Figure 11.2.

Many people are confused when they see the calutron photo because they thought that Iraq was *not* developing nuclear weapons. Such people are confusing two time frames. Back in 1990, Iraq certainly was attempting to develop nuclear weapons. This fact led many analysts to conclude, incorrectly, that it was doing so again in 2002, just before the United States invaded for the second time.

Figure 11.2. An Iraqi calutron after being destroyed by the United Nations.

How Worried Should We Be?

A good friend of mine, a nuclear expert who designed some of the early atomic bombs, is frighteningly pessimistic. He predicts that a nuclear weapon will be exploded in anger sometime in the next 10 years, with about a 50% chance. The biggest threat no longer comes from Russia or (as some think) the United States. Israel might use its putative weapons if it perceives that its existence is threatened. With the continuing conflict between India and Pakistan over Kashmir, war between those countries could erupt at any time, and they both have nukes. Then there is the terrorist threat. When the Soviet Union collapsed, its accounting system for its large arsenal of weapons collapsed too. It is not clear whether any of its weapons have been diverted or sold, perhaps to terrorists.

Is the danger as high as my friend warns? Some believe the dangers of a terrorist bomb are exaggerated, and with the end of the Soviet threat we should all rejoice, if not relax. Why do opinions differ so much? I think the reason is that the nuclear experts are not experts in world affairs, and the politicians and diplomats are not experts in nuclear matters. Really nobody is in a position to evaluate everything needed to reach a definitive conclusion. Personally, I am far more worried about future biological attacks than I am about nuclear ones. That thought might not give you much comfort.

Nukular Weapons

The Hiroshima bomb was announced by President Harry Truman in 1945. He referred to it as an *atomic bomb*. This name is still used. Is it a valid name? Some say no, arguing that ordinary TNT uses the energy of the combining and breaking up of atoms, so *atomic* should be reserved for ordinary explosives.

What do you think? Is *atom bomb* a legitimate name? Do we need a different term for a weapon that uses the energy of the nucleus? Are names determined by strict physics? On the subject of English usage, I speak with no more authority than you. My general feeling is that the term *atomic bomb* is just as good as any other. The nucleus is, after all, part of the atom. In some sense, all previous weapons were molecular bombs, in that they released energy from the binding between atoms. The nuclear weapons of World War II were the first to release the energy buried in individual atoms.

What about "nukular"? That's the way the word *nuclear* has been pronounced by many people, including President Dwight Eisenhower, Edward Teller, President Jimmy Carter (who served on a nuclear submarine), and more recently, President George W. Bush. Is this pronunciation wrong? Many scholars assume that it is. Academics mock people who speak that way. Yet "nukular" is included as a colloquial variant in both the Merriam-Webster and Oxford English dictionaries. Edward Teller was one of the inventors of the hydrogen bomb; he is often called the "father" of the hydrogen bomb. Doesn't he have the right to say it any way he wants?

It turns out that pronouncing the word as "nukular" has been a tradition at some of our weapons laboratories since World War II. Yes, I know, it isn't spelled that way. I personally pronounce it as it's spelled. Some scholars suggest that the pronunciation "nukular" derives from the combination of *nuke* with the ending *-ular* inspired by similar words, such as *spectacular*, *popular*, and *molecular*. Is it really wrong? After all, we have lots of words in the English language that are not pronounced the way they are spelled. Again, this is not my subject of expertise. See the interesting article in Wikipedia (http://wikipedia.org) under the entry *nucular*. But be prepared: if you follow the lead of Edward Teller but don't have his credentials, people will mock you if you say it the way he did. That, too, may be important for a future president to know.

12

Nuclear Power

NUCLEAR POWER. What is your reaction to those words? In a 1997 *Frontline* documentary, "Nuclear Reactions," a number of citizens were asked to write down the words they associated with that phrase. Their answers were *disaster, trouble, disgusting, dangerous,* and *radiation.* Yet, in a 2005 op-ed piece in the *New York Times,* the "liberal" columnist Nicholas Kristof wrote that "Nukes Are Green." How could he conclude that, after thousands were killed at the Chernobyl disaster, when everyone knows that the nuclear waste problem is unsolved? What's going on? Should you be *for* nuclear power or *against* it?

There is great confusion on this question not only in the minds of the public but in those of our leaders. Many people on both sides of this divisive issue think that their point of view is obvious, and that makes them suspicious of those who disagree. Nuclear power is a problem that future presidents will have to contend with, not only in making decisions, but in convincing the public that their decisions are correct. To accomplish these difficult tasks, you will need to understand the physics.

The most famous symbol of nuclear power reactors is the building shown in Figure 12.1. The tall, gracefully curved structure is not the

Figure 12.1. A nuclear power plant, with the cooling tower in prominent view.

nuclear reactor, but only a cooling tower. The reactor itself is inside the small vertical cylindrical structure just behind the tower and to its left. In this photo, the reactor building was still under construction; it had not yet received its semi-spherical cap.

The physics of the nuclear reactor is very similar to that of nuclear bombs. In both, the energy is released from a chain reaction of uranium-235 and/or plutonium-239. The bomb uses doubling to get all the fission to take place in a millionth of a second. In contrast, the reactor uses a *sustained* chain reaction, with no doubling. In the reactor, fission releases two or three neutrons, but on average only one of those goes on to hit another nucleus and trigger another fission, so the rate of reaction doesn't grow. Energy is released at a constant rate, heats water, and produces steam, which in turn runs a turbine that runs a generator that produces electricity. Gas and coal power plants work in a similar way, except that they use fossil fuels rather than uranium to produce the heat. A large nuclear plant produces about as much electric power as a large gas or coal plant: a billion watts of electricity, called a *gigawatt*. Since a typical home uses 1000 watts, the output of one of these reactors can supply power to a million homes. About 20% of the electricity generated in the United States comes from its 104 nuclear power plants.

Some people think it is strange that nuclear power is used simply to heat water. Someday we may be able to use fission to create electricity directly, but right now, the hot-water approach is the most practical. That's true for nuclear submarines too. It is amusing to think of a nuclear submarine being run by steam, but that's the way it works. The fundamental difference between the old diesel submarine and the new nuclear one is that the nuclear fuel has over a million times the energy content (for the same weight), and the nuclear sub doesn't require oxygen, so it can stay at depth for extended periods. In fact, some of the energy from the nuclear reactor is used to separate oxygen out of water to provide artificial air for the crew. The hydrogen that comes out of the H_2O along with the oxygen is typically bubbled off into the ocean as waste. It can't be burned without using the valuable oxygen that was released in the same process.

Look at the diagram of a nuclear power plant shown in Figure 12.2. It may seem complicated, but it is worth examining. The only part that is actually nuclear is the reactor, which is encased in a thick metal vessel, which in turn is inside a reinforced concrete building. The purpose of these containers within containers is to prevent radioactivity from escaping into the environment if the fuel or waste is accidentally released from the pellets in the reactor. The water from the reactor vessel circulates to the steam generator, where water is boiled. The steam goes to the turbine, which operates the generator to make electricity.

Although both a nuclear bomb and a nuclear reactor can use ura-

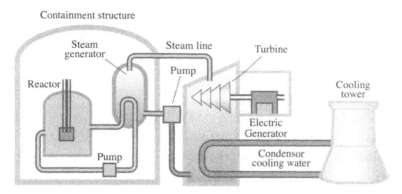

Figure 12.2. Schematic of a nuclear power plant. The sustained chain reaction takes place in the small reactor. The heat from the fission is used to make steam; the steam runs a turbine that operates a generator to make electricity.

nium fuel, the purity required is vastly different. Recall that natural uranium is 99.3% uranium-238 and only 0.7% uranium-235. For the chain reaction to proceed in the bomb, the uranium-238 contaminant must be removed so that it doesn't absorb too many neutrons. Reactors are different. The first reactor ever built, by Enrico Fermi during the World War II Manhattan Project, used natural unenriched uranium, with 99.3% contaminant. How can that be? Why was he able to use non-enriched natural uranium in a reactor?

The answer is based on a key discovery about the peculiar nuclear physics of uranium-238: although it has a very high absorption rate for the fast neutrons emitted in fission, it has only a very low rate for slow ones. If you slow the neutrons down before they hit the uranium-238, most simply bounce off. To accomplish this slowing, Fermi added a material to the reactor that he called a *moderator*. His moderator was graphite, a form of carbon. The uranium was put in rods small enough that most of the fast neutrons leaked out and entered the carbon. There they bounced off the relatively cool carbon atoms, losing energy with each bounce, and finally (by chance) wandering back into the uranium pellets. By then they were slow enough that fewer than half of them were absorbed on the uranium-238. The reactor was designed so that, on average, one neutron from each fission would come back and find a uranium-235 nucleus and trigger a new fission. Thus the reaction was sustained. The slowed neutrons are sometimes referred to as *thermal neutrons* because after several bounces they have no more energy than they would have from the temperature of the reactor.

Everything in these designs is based on the peculiar properties of the various nuclei. Carbon doesn't absorb most of the fast neutrons; uranium-238 doesn't absorb most of the slow ones. As future president, you will not need to understand nuclear structure; all you will really have to know is that moderating the neutrons—slowing them down—means you don't have to enrich the uranium. Moreover, the moderation will have enormously important consequences for the safety of the reactor, as we'll discuss in the next section.

Graphite was also the moderator used in the Chernobyl reactor. The moderator used in Canada is heavy water—deuterium oxide, D_2O. In the United States, we use ordinary water, H_2O. To emphasize the contrast with heavy water, this water is sometimes called *light water*. Only in the nuclear industry is ordinary water given a special name.

It turns out that ordinary water does absorb some neutrons. They tend to stick to the hydrogen, turning it into deuterium. That means that enough neutrons are lost to prevent the chain reaction from continuing. To get around this, the US reactors use partially enriched uranium, consisting of about 3% uranium-235. This is often referred to as *reactor-grade* uranium, even though completely natural uranium will work in reactors moderated by carbon or by heavy water. Most of our uranium enrichment plants are manufacturing reactor-grade uranium, not bomb-grade. It is important to know that reactor-grade uranium cannot be used to make a nuclear weapon.

Can a Nuclear Reactor Explode Like an Atomic Bomb?

What if we lose control of the chain reaction? Isn't every nuclear reactor simply a nuclear bomb waiting to explode? Are we depending on the vigilance of the reactor operator to prevent such an explosion? Isn't this an extremely dangerous technology that we should avoid at all costs? The answer is obvious to most members of the public: yes—so nuclear power is not worth the risk. Fear of nuclear disaster is likely what triggered the strong word associations mentioned at the beginning of this chapter. The fear has been amplified by fictional depictions of nuclear power catastrophes, as in the 1979 movie *The China Syndrome*, and by the fact that the most famous nuclear reactor safety inspector, watched by millions on TV, is named Homer Simpson.

Here is a key fact for future presidents: a nuclear reactor cannot explode like an atomic bomb. The public image is not correct. The reason that I can confidently make this statement has nothing to do with control; it is based in indisputable physics. No expert disagrees. There are other dangers from nuclear power reactors, but exploding like an atomic bomb is not one of them. Of course, a future president must know more than the answer. You must be ready and able to explain and defend what you say and to convince the public that you really do know what you're talking about, that you're not just a dupe of a nuclear industry that supported your campaign.

The physics reason for the safety of the nuclear power reactor is the fact that the presence of lots of uranium-238 in the core means that the nuclear chain reaction will stop if the neutrons are not moderated. I repeat: unless the neutrons are slowed down, they will be absorbed on the uranium-238 and the chain reaction will cease.

Can a runaway chain reaction happen with moderated neutrons? Yes; in fact, that is what happened in the Chernobyl accident in 1986. We'll discuss this in more detail near the end of the chapter. But with moderated neutrons, the energy released is severely limited because the explosion breaks up the reactor before much energy can be released. Basically the explosion outruns the neutrons. That does not happen in a nuclear bomb, because the bomb uses pure uranium-235, so the chain reaction can proceed using fast neutrons.

Think of it this way: if a moderated nuclear reactor runs out of control, the energy from fission will build until the energy density is enough to blow the structure apart. That typically happens when the energy density is comparable to that of TNT. At Chernobyl, the energy released was enough to destroy the building and start the carbon moderator burning. The smoke and heat carried much of the radioactive waste (the fission fragments that had been produced during the reactor operation) into the atmosphere, where they dispersed over the landscape.

Nuclear reactors can explode—but like dynamite, not like an atomic bomb. Such an explosion is worse than a dynamite explosion because of the radioactive waste that it releases, not because of the blast. The result would be different if we used weapons-grade uranium or plutonium as the fuel; then it could explode like an atomic bomb. There is talk of using such fuel for a device called a *fast breeder reactor*. I'll come to that after I discuss the manufacture of plutonium.

Manufacturing Plutonium

In a nuclear reactor, only one of the two neutrons from uranium fission is used to trigger another fission. The other is absorbed. Some of the neutrons are absorbed on *control rods* made of material such as boron or cadmium; these elements absorb neutrons on their nuclei without fissioning. The control rods can be moved in and out of the core to control the rate of fission. Some of the neutrons are absorbed by the moderator; in fact, the reason that heavy water is more effective than light water is that it is less likely to grab neutrons in this way. Some of the neutrons are absorbed on the uranium-238; recall that this form constitutes 97% of the uranium in US reactors.

When uranium-238 absorbs a slow neutron, it doesn't fission but instead becomes uranium-239. (The number refers to the total number of protons and neutrons in the nucleus; absorb an extra neutron, and this number increases by one.) Uranium-239 is radioactive and decays (it emits an electron and a neutrino, with a half-life of about 23 minutes), turning into the element neptunium. Neptunium is also radioactive, with a half-life of 2.3 days, and when it decays it turns into the famous form of plutonium—the one that can be used for a nuclear weapon: plutonium-239. As long as the nuclear reactor is operating, it will continue to produce—we might even say *manufacture*—plutonium. Of course, what is really happening is that the reactor is converting the abundant form of uranium, uranium-238, into plutonium.

After the reactor has run for a year or more, much of the fissionable uranium-235 has been depleted. Plutonium has been produced, and it fissions too, helping in the chain reaction. In a normal power reactor, the produced plutonium is not enough to replace the used-up uranium-235, so the waste must be removed and the fuel replaced. When this depleted material is removed from the reactor, it has plutonium mixed in with it. This is potentially valuable material because it, too, can undergo a sustained chain reaction. If separated from the rest of the waste, it could be used as a nuclear fuel itself in a bomb or another reactor. It is tricky, however, to remove the plutonium because the waste is full of highly radioactive fission fragments. But since plutonium is a distinct element, it has chemical properties that allow it to be separated. Doing chemistry with highly radioactive materials is called *radiochemistry*, and it is a highly developed art.

Methods for extracting plutonium from waste were published during the Atoms for Peace program in the 1950s, when the United States shared its nuclear technology with the world in exchange for promises that other countries would forgo nuclear weapons. The extraction process is called *reprocessing*, and it is frequently in the news. North Korea reprocessed its fuel, violating its prior agreement not to do so, and obtained plutonium for its bomb in that way. France reprocesses its spent fuel to reuse it in reactors. The United States does not, for reasons we will discuss. Instead, the plutonium is included as part of the nuclear waste. The fact that the waste contains this long-lived radioactive material (having a half-life of 24,000 years) with such a fearsome reputation (it is used in nuclear bombs) has led to a public relations problem for the US Department of Energy. The next chapter is devoted to nuclear waste disposal.

Fast Breeder Reactors

In France, plutonium-239 is not considered nuclear waste, because it can be used as a nuclear fuel to run a nuclear reactor. If you put some uranium-238 in the reactor, some of the fission neutrons will generate more plutonium. Can you wind up with more plutonium than you started with? The surprising answer is yes! The reason is that plutonium fission yields three neutrons, so if one goes to the next fission, you have two left over to make more plutonium. There are other tricks too. When fast neutrons hit lithium, they release additional neutrons; each of these can, in principle, turn a uranium-238 nucleus into a plutonium nucleus. Uranium-238 is abundant. Natural uranium contains 140 times more of this form than of the fissionable uranium-235.

A reactor that uses plutonium fission to produce more fuel than it consumes is called a *breeder* reactor. In the breeder reactors that have been designed, it would take about 10 years to double the fuel. Breeder reactors have faced public opposition, however. The two most common objections are the dangers of a "plutonium economy," and the real possibility of a large nuclear explosion.

The phrase *plutonium economy* refers to a world in which plutonium is a common fuel, used not in hundreds of reactors, but perhaps thousands. That scenario increases the risk that some might be diverted by terrorists or rogue states to make nuclear weapons. Proponents respond that the dangers of plutonium have been greatly exaggerated, when in reality plutonium bombs are extremely difficult to make because they require an exquisitely tuned implosion. Opponents counter that even a fizzled plutonium bomb would spread deadly plutonium around a city. Proponents reply that such dirty bombs are not really much of a threat, because dispersed radioactivity is not very lethal; we live with radioactivity all the time. The debate continues, with much passion on both sides. As a future president, you will have to fold into all these risks the other dangers of nuclear power, the need for energy independence, and the global-warming risks of fossil fuel.

There is another argument against breeding plutonium. The most efficient kind of breeder reactor would use fast, not slow, neutrons, and for that reason it is called a *fast breeder*. But if fast neutrons are used, then the intrinsic physics safety of the reactor is lost. In a fast breeder the chain reaction could spread uncontrollably, and instead of just a meltdown, the reactor really could explode like an atomic bomb. Proponents respond that the design and geometry of the system would prevent this from happening. Even so, a future president should know that not *all* nuclear reactor designs are protected from explosion by fundamental physics.

The China Syndrome

The term *China Syndrome* was coined by someone with a strange sense of humor. It is a name for the worst possible nuclear reactor accident that analysts could imagine for the current design of US nuclear power reactors. This hypothetical disaster begins with a meltdown of the fuel. Recall that the main purpose of a nuclear power reactor is to produce heat to make steam to drive a turbine to make electricity. Fission produces the heat, and water carries the heat away. Suppose this water is lost, perhaps because of a pipe leak. What happens next?

One thing that happens immediately is surprising to most people: the chain reaction stops. That's because the cooling water is also a moderator; it slows neutrons. When the water is gone, the neutrons are not moderated. They remain fast (the way they were when emitted), and because of the peculiar nuclear properties of uranium-238, many of them are absorbed on that nucleus—one that does not emit enough neutrons to give a sustained chain reaction. The chain reaction stops. Immediately.

Despite the cessation of the chain reaction, a lot of radioactivity remains in the core from the fission fragments—enough that, without the cooling water, the reactor core gets hotter and hotter. In a typical reactor, this situation leads to the injection of emergency cooling water. All US reactors are required to have such a system.

When an explosion occurred at the Chernobyl nuclear reactor, the Russians announced that the chain reaction had stopped—in that case, because of damage to the core. The chairman of the US Senate Intelligence Committee announced on television that this was a "blatant lie." I cringed. He was confusing the chain reaction with the decay of the remaining fission fragments. He knew the radioactivity hadn't stopped, but he didn't realize that the Soviets were being completely honest. The fact that the chain reaction had stopped was important; it meant that the level of power being produced had dropped enormously. Please remember this when you become president.

What if the emergency core cooling system fails? (Remember, we are trying to imagine the *worst* possible accident.) If that happens, the fuel will continue to get hot from residual radioactivity of the fission fragments until it finally melts. The heat will also melt the walls of the capsules that contain the fuel, and the hot liquid fuel will dribble down to the base of the steel reactor vessel and form a puddle. The radioactivity will drop rapidly as the nuclei that decay are used up. Even so, calculations show that the fuel puddle might continue to get hotter, perhaps even hot enough to melt the steel of the reactor vessel. If it does melt through, the radioactive material will begin to damage the concrete floor of the containment building. That floor is several feet thick, but can we be sure the hot fuel won't melt all the way through it too? If it does, the radioactivity will be outside. Volatile gases in the waste might then be able to escape into the atmosphere. The hot liquid fuel may continue to melt its way down. If it stays concentrated, then it will just keep going—all the way "to China."

No, obviously it won't reach China. (Besides, China isn't on the other side of the Earth of any of our nuclear reactors). The fuel is unlikely to get very far, because it spreads out and that allows it to cool. That's little consolation, however. The containment has been breached. Radioactive gases and volatile elements such as iodine are the biggest worry. They are what caused the most cancer from Chernobyl.

The radioactivity in a nuclear power plant is huge, enough to kill 50 million people if they ate it. Even a small amount leaked into the atmosphere can do great damage. Recall that the best estimate (based on the linear hypothesis) is that cancer induced from the release of radioactivity at the Chernobyl accident will kill 4000 people. The United States had a nuclear accident too, at Three Mile Island.

Three Mile Island—Fuel Meltdown

On March 29, 1979, a nuclear reactor in the United States ran out of control and a third of its uranium fuel melted. It happened in the power plant on Three Mile Island near Harrisburg, Pennsylvania. The incident began when pumps that fed external cooling water to the reactor failed. A backup pump had inadvertently been left with a critical valve closed. Control rods were immediately plunged into the reactor core, stopping the chain reaction, but energy from the decaying fission fragments continued to heat the core. Other safety systems failed because of poor design or human error. A technician turned off an emergency core cooling system because he mistakenly thought the reactor was full of water. As the uranium overheated, the fuel and waste products turned to liquid. The radioactive material did not melt through the steel containment vessel that held the fuel, so the rest of the China Syndrome did not take place. However, some of the water that cooled the fuel leaked into the concrete containment building, and radioactive gases dissolved in this water made the interior of the building very radioactive. In order to prevent pressure from building up, some of this gas was purposely leaked to the outside environment by the operators. Calculations show that the number of expected cancers from this leakage (assuming the linear hypothesis) was about one.

This accident happened right after the movie *The China Syndrome* was released. Many people mistakenly thought the accident was as terrible as the one in the movie. Others knew it was not as bad, but feared that it was *potentially* as bad.

After the accident at Three Mile Island, many people who lived nearby bought Geiger counters and measured the radioactivity in the region. They discovered it was high—typically 30% above the national average. This finding caused enormous concern, although it puzzled the experts because the radioactivity released from the plant was far too small to account for it. It was finally determined that the high level was characteristic of that area, and had been that way long before the accident. It comes from uranium in the local soil, which decays to radioactive radon gas. For the 50,000 people who lived near Three Mile Island, such natural radioactivity would lead to 60 excess deaths.[36]

In the *Frontline* documentary "Nuclear Reactions," a resident near the Three Mile Island reactor was convinced of great harm from the release of radioactivity from the accident. When she was asked about the much larger dangers from ground radon, she responded that she had no fear of that because it was "natural." She didn't appreciate that human cells can't distinguish between natural and artificial radiation.

The Three Mile Island accident is vivid in the minds of everyone who lived at the time. Much of what they remember is fragmented or confused. As a future president, you will need to remember the details accurately. The accident was terrible, but not as bad as most people remember. The real issue is how close it came to being worse. That is hard to know, but it is important to recognize that although the fuel did melt and the emergency core cooling system failed, the fuel did not melt through either the steel reactor vessel or the concrete containment building.

Chernobyl—A Runaway Chain Reaction

The worst nuclear reactor accident of all time occurred near the village of Chernobyl (in Ukraine) on April 26, 1986. We discussed the consequences of this event previously, when we talked about the linear hypothesis. The Chernobyl reactor used carbon for a moderator—as did the first reactor built by Enrico Fermi during World War II—unlike US reactors, which use water. During an experiment designed to test the safety of the reactor, the chain reaction began to grow out of control. This result was due in part to operator error, and in part to bad design. The Chernobyl reactor had a *positive temperature coefficient*. That meant that as the reactor heated, it tended to increase the rate of the chain reaction. As a result, the reactor got so hot that the cooling water explosively boiled, in what is called a *steam explosion*. Without cooling, the carbon began to burn, and much of the radioactivity in the core (from the fission fragments) spread out from the reactor carried by smoke. It is estimated that 5% to 30% of the fission fragments in the core spread to the surrounding countryside. These weren't just the gases and volatiles, but even the solid components. In that sense, this accident was much worse than the "worst-case scenario" China Syndrome possible for a US-style water-moderated reactor. For US plants the reactivity coefficient is negative, so overheating tends to turn off the chain reaction. Figure 12.3 shows the damaged power plant.

Incredibly (to US scientists and engineers), the Chernobyl plant didn't even have a containment building. If it had had one, the accident may very well have caused virtually no deaths. The Chernobyl accident was not an example of the China Syndrome—a meltdown from loss of coolant. It was a *reactivity* accident from a runaway nuclear chain reaction. Because the chain reaction depends on slow neutrons, it shut off as soon as the temperature reached a point to cause a small explosion. The subsequent fire and smoke spread most of the radioactivity.

Figure 12.3. The destroyed reactor at Chernobyl. Unlike US reactors, this one had no concrete containment building.

The UN estimates 4000 excess cancer deaths from Chernobyl. Is this the worst industrial accident ever? No; nuclear reactors are not unique in their ability to kill. In 1984, a chemical plant leak near Bhopal, India, killed 5000 people immediately, and some estimate the ultimate toll will reach 20,000.

Could a truly safe nuclear reactor be designed? Some people say that the US design is already safe enough. But can one be designed that is so safe that we would not have to worry about either a reactivity accident or a loss-of-coolant accident? The answer is yes. One such design is called a *pebble bed reactor*.

Pebble Bed Reactors

In a pebble bed reactor design, the uranium is placed in a 2½-inch-deep bed of pebbles of pyrolytic graphite—a type of graphite that is very resistant to high heat, unlike the graphite used in Chernobyl. The pebbles are covered with equally tough silicon carbide ceramic shells. The materials are sturdy under the highest temperatures that the reactor can reach, even if everything goes wrong.

The safety of the pebble bed reactor is based on physics. At higher and higher temperatures, the neutrons from fission come out faster and faster, just because of the heat motion of the uranium nuclei. But if they get too fast (because the reactor is too hot), they reach the energy at which they are absorbed by the non–chain reaction component of the uranium, the uranium-238. These nuclei absorb *only* high-energy neutrons, and although they fission, they do not release enough new neutrons to sustain the chain reaction. As a result, if the reactor exceeds its operating temperature, the chain reaction actually slows down. The temperature at which this happens is too low to damage the materials that make up the pebbles, so the reactor is intrinsically safe against meltdown, explosion, and fire. No human intervention is required; no mechanical system has to turn on. The reactor is designed such that if anything (or even everything) electrical, mechanical, and fluid fails, it will automatically slow down, thanks to physics, and the reactor will revert to a safe "idling" temperature. You can even remove the control rods and the coolant (usually helium gas), and nothing bad happens.

Of course, fission fragments are still produced, so after a few months of normal operation the pebbles build up a lot of radioactivity. Within each pebble the nuclear fuel is subdivided into smaller *seeds* that make containment even more effective. If a pebble is damaged—for example, when it is being removed—some radioactivity may leak out, but only the radioactivity in that single pebble, and usually only the part of the radioactivity that is in gaseous form. That happened in Germany in 1986, leading to the shutdown of a pebble bed facility in that country—considered by many to be an overreaction due to nuclear fear. For waste storage, the nuclear waste never has to be removed from the spent pebbles. In every way that I can imagine, pebble beds are safer than the present nuclear reactors, and arguably safer than the global-warming dangers posed by fossil fuels.

Pebble bed reactors are designed to be used in modules. That means that if you want more power, instead of making the reactor bigger or running it hotter, you just add more modules. For that reason, you'll sometimes hear such a reactor referred to as a *pebble bed modular reactor*, or *PBMR*.

The high temperature at which the pebble bed reactors work leads to higher efficiency. Instead of 32% to 35% of the energy from the uranium going into electricity, with the pebble bed reactor that range is 40% to 50%. That is an economic advantage: more useful power from the same fuel.

Pebble bed reactors do suffer from a bureaucratic problem: the nuclear reactor rules written for the older generation of reactors may not be appropriate for the newer designs, yet they are the law. For example, some rules mandate emergency core cooling systems for all power reactors, yet such a system does not make sense for a core than cannot overheat. When you become president, you may want to review and update some of these rules.

13

NUCLEAR WASTE

AS PEOPLE recognize the dangers of fossil fuel plants—especially the risk of global warming from carbon dioxide production—nuclear power begins to look more attractive. But what about the waste—all that highly radioactive debris that will endure for thousands of years? Do we have the right to leave such a legacy to our children?

Nuclear waste is one of the biggest technical issues that any future president is likely to face. The problem seems totally intractable. Plutonium—just one of the many highly radioactive waste products—has a half-life of 24,000 years. Even in that unimaginable amount of time, its intense radioactivity will decrease by only half. After 48,000 years it will still emit deadly radiation at a quarter of its original level. Even after 100,000 years the radiation will still be above 10% of the level it had when it left the reactor. What if it leaks into the ground and reaches human water supplies? How can we possibly certify that this material can be kept safe for 100,000 years?

Still, the US government persists in its pursuit of "safe" nuclear waste disposal. It has created a prototype nuclear waste facility buried deep within Yucca Mountain, Nevada (Figure 13.1). To keep the waste safe, the storage rooms are 1000 feet below the surface. To

Figure 13.1. Yucca Mountain, Nevada, the site of the prototype nuclear waste storage facility.

store even part of the present nuclear waste requires a vast area, nearly 2 square miles. The cost of the facility is expected to reach $100 billion, with hundreds of billions of dollars more in operating costs.

To make matters worse, the Yucca Mountain region is seismically active. More than 600 earthquakes of magnitude 2.5 and higher have occurred within 50 miles in the last decade alone. Moreover, the region was created by volcanic activity. Although that was millions of years ago, how sure can we be that the waste facility won't be torn apart by another eruption?

Many alternatives have been suggested for nuclear waste storage. Why not just send the waste into the sun? Well, maybe that's not such a good idea, since on launch some rockets do crash back down on the Earth. Some scientists have proposed that the waste be put in vessels and sunk under the oceans, in a region where the movement of the Earth's crustal plates will subduct the material, eventually burying it hundreds of miles deep. Yet just the fact that scientists make such suggestions seems to emphasize how severe the problem really is.

Here is the worst part. We have already generated more than enough nuclear waste to fill up Yucca Mountain. That waste won't go away. Yet you, a future president, are considering *more* nuclear power? Are you insane?

My Confession

The furor against nuclear power has been so intense that I felt compelled to reproduce the anti-nuke viewpoint in the opening of this chapter, including at least part of their passion. These are the arguments that you will hear when you are president. Yet it hardly matters whether you are pro-nuke or anti-nuke. The waste is there, and you will have to do something with it. You can't ignore this issue, and to do the right thing (and to convince the public that you're doing the right thing) you must understand the physics.

When I work out the numbers, I find the dangers of storing our waste at Yucca Mountain to be small compared to the dangers of not doing so, and significantly smaller than many other dangers we ignore. Yet the contentious debate continues. More research is demanded, but every bit of additional research seems to raise new questions that exacerbate the public's fear and distrust. I have titled this section "My Confession" because I find it hard to stand aside and present the physics without giving my own personal evaluation. Through most of this book I've tried to present the facts, and just the facts, and let you draw the conclusions. In this section, I confess that I'll depart from that approach. I can't be evenhanded, because the facts seem to point strongly toward a particular conclusion.

I've discussed Yucca Mountain with scientists, politicians, and many concerned citizens. Most of the politicians believe the matter to be a scientific issue, and most of the scientists think it is political. Both are in favor of more research—scientists because that is what they do, and politicians because they think the research will answer the key questions. I don't think it will.

Here are some pertinent facts. The underground tunnels at Yucca Mountain are designed to hold 77,000 tons of high-level nuclear waste. Initially, the most dangerous part of this waste is not plutonium, but fission fragments such as strontium-90, an unstable nucleus created when the uranium nucleus splits. Because these fission fragments have shorter half-lives than uranium, the waste is about 1000 times more radioactive than the original ore. It takes 10,000 years for the waste (not including plutonium, which is also produced in the reactor, and which I'll discuss later) to decay back to the radioactive level of the mined uranium. Largely on the basis of this number, people have searched for a site that will remain secure for 10,000 years. After that, we are better off than if we left the uranium in the ground, so 10,000 years of safety is probably good enough, not the 100,000 years that I mentioned in the chapter introduction.

Ten thousand years still seems impossibly long. What will the world be like 10,000 years from now? Think backward to appreciate the amount of time involved: Ten thousand years ago humans had just discovered agriculture. Writing wouldn't be invented for another 5000 years. Can we really plan 10,000 years into the future? Of course we can't. We have no idea what the world will be like then. There is no way we can claim that we will be able to store nuclear waste for 10,000 years. Any plan to do that is clearly unacceptable.

Of course, calling storage unacceptable is itself an unacceptable answer. We have the waste, and we have to do something with it. But the problem isn't really as hard as I just portrayed it. We don't need absolute security for 10,000 years. A more reasonable goal is to reduce the risk of leakage to 0.1%—that is, to one chance in a thousand. Because the radioactivity is only 1000 times worse than that of the uranium we removed from the ground, the net risk (probability multiplied by danger) is $1000 \times 0.001 = 1$—that is, basically the same as the risk if we hadn't mined the uranium in the first place. (I am assuming the linear hypothesis—that total cancer risk is independent of individual doses or dose rate—but my argument won't depend strongly on its validity.)

Moreover, we don't need this 0.1% level of security for the full 10,000 years. After 300 years, the fission fragment radioactivity will have decreased by a factor of 10; it will be only 100 times as great as the mined uranium. So by then, we no longer need the risk to be at the 0.1% level, but could allow a 1% chance that all of the waste leaks out. That's a lot easier than guaranteeing absolute containment for 10,000 years. Moreover, this calculation assumes that 100% of the waste escapes. For leakage of 1% of the waste, we can accept a 100% probability after 300 years. When you think about it this way, the storage problem begins to seem tractable.

However, the public discussion doesn't take into account these numbers, or the fact that the initial mining actually removed radioactivity from the ground. Instead, the public insists on absolute security. The Department of Energy continues to search Yucca Mountain for unknown earthquake faults, and many people assume that the acceptability of the facility depends on the absence of any such faults. They believe that the discovery of a new fault will rule Yucca Mountain out. The issue, though, should not be whether there will be any earthquakes in the next 10,000 years, but whether after 300 years there will be a 1% chance of a sufficiently large earthquake that 100% of the waste will escape its glass capsules and reach groundwater. Or, we could accept a 100% chance that 1% of the waste will leak, or a 10% chance that 10% will leak. Any of these options leads to a lower risk than if the original uranium had been left in the ground, mixing its natural radioactivity with groundwater. Absolute security is an unnecessarily extreme goal, since even the original uranium in the ground didn't provide it.

The problem is even easier to solve when we ask why we are comparing the danger of waste storage only to the danger of the uranium originally mined. Why not compare it to the larger danger of the natural uranium left in the soil? Colorado, where much of the uranium is obtained, is a geologically active region, full of faults and fissures and mountains rising out of the prairie, and its surface rock contains about a billion tons of uranium.[37] The radioactivity in this uranium is 20 times greater than the legal limit for Yucca Mountain, and it will take more than 13 billion years—not just a few hundred—for the radioactivity to drop by a factor of 10. Yet water that runs through, around, and over this radioactive rock is the source of the Colorado River, which is used for drinking water in much of the West, including Los Angeles and San Diego. And unlike the glass pellets that store the waste in Yucca Mountain, most of the uranium in the Colorado ground is water-soluble. Here is the absurd-sounding conclusion: if the Yucca Mountain facility were at full capacity and all the waste leaked out of its glass containment immediately and managed to reach groundwater, the danger would still be 20 times less than that currently posed by natural uranium leaching into the Colorado River. The situation brings to mind the resident near Three Mile Island who feared the tiny leakage from the reactor but not the much greater radioactivity of natural radon gas seeping up from the ground.

I don't mean to imply that waste from Yucca Mountain is not dangerous. Nor am I suggesting that we should panic about radioactivity in the Los Angeles water supply. The Colorado River example illustrates only that when we worry about mysterious and unfamiliar dangers, we sometimes lose perspective. Every way I do the calculation, I reach the same conclusion: waste leakage from Yucca Mountain is not a great danger. Put the waste in glass pellets in a reasonably stable geologic formation, and start worrying about real threats—such as the dangers of the continued burning of fossil fuels. I'll discuss that in the final part of this book.

A related issue is the risk of mishaps and attacks during the transportation of nuclear waste to the Yucca Mountain site. The present plans call for the waste to be carried in thick, reinforced concrete cylinders that can survive high-speed crashes without leaking. In fact, it would be very hard for a terrorist to open the containers, or to use the waste in radiological weapons. The smart terrorist is more likely to hijack a tanker truck full of gasoline, chlorine, or another common toxic material and then blow it up in a city. Recall from the chapter on terrorist nukes that al-Qaeda told José Padilla to abandon his effort to make a dirty bomb and instead focus his efforts on natural-gas explosions in apartment buildings.

Why are we worrying about transporting nuclear waste? Ironically, we have gone to such lengths to ensure the safety of the transport that the public thinks the danger is greater than it really is. Images on evening newscasts of concrete containers being dropped from five-story buildings, smashing into the ground and bouncing undamaged, do not reassure the public. This is a consequence of the "where there's smoke there's fire" paradox of public safety. Raise the standards, increase the safety, do more research, study the problem in greater depth, and in the process you will improve safety and frighten the public. After all, would scientists work so hard if the threat weren't real? Scientists who propose rocketing the waste to the sun, or burying it in a subduction zone in the ocean, also seem to be suggesting that the problem is truly intractable, and that premise exacerbates the public fear.

Plutonium

Let me return now to the danger of the plutonium in the waste. Plutonium is not a fission fragment; it is produced in the reactor when uranium absorbs neutrons. Unlike the fission fragments, plutonium doesn't go away by a factor of 10 in 300 years; its half-life is 24,000 years. Not only that, but many people think plutonium is the most dangerous material known to man.

Is plutonium really more dangerous because of its long half-life? Coal-burning plants bury their waste in the ground. They are not very radioactive, but the ashes are high in carcinogens. What if these ashes get into the groundwater? How safe is coal, as an alternative to nuclear resources? These carcinogens don't disappear with a half-life of 24,000 years, like plutonium. They last essentially forever, unless they decompose or are dissolved in groundwater. If we take lifetime in the ground to be a measure of danger, then the carcinogens of coal are far worse than plutonium.

Plutonium is certainly dangerous if you make nuclear weapons out of it. If turned into an aerosol and inhaled, it is more toxic than anthrax—and that's very toxic. Yet it is still not a record setter; botulism toxin (the active ingredient in Botox) is a thousand times worse. But for nuclear waste, people worry about the plutonium dissolved in groundwater. Such plutonium is toxic, but not worse than other threats. The cancer dose for such water requires that you drink enough to consume ½ gram of the dissolved plutonium.[38] Plutonium is about five times *less* toxic than cyanide. Moreover, plutonium is not very soluble; when mixed with water typically only one part in 100,000 actually dissolves. The horrendous threat from ingested plutonium is an urban legend—believed to be true by many people, yet false. It gains credibility every time a prominent person inadvertently passes it on, like a computer virus. Even the TV anchorman who was once the most trusted man in America, Walter Cronkite, sententiously (but incorrectly) pronounced plutonium to the be the deadliest material known to man on *The CBS Evening News.*[39]

It is probably a mistake to bury the plutonium with the waste. It is a good fuel for reactors, as valuable as uranium. Part of the original reason for burying it (rather than extracting and using it) was to keep the public from worrying about it, but that approach has backfired. A second reason for avoiding the reprocessing is that it is currently cheaper to mine fresh uranium and enrich it than it is to extract the plutonium. The full cost of not extracting the plutonium, however, must include the enormously increased expense of storing it at Yucca Mountain—an expense that was once considered small but has now grown.

By any reasonable measure I can find, the Yucca Mountain facility is safe enough. It is far safer to put the waste there than to leave it on-site at the nuclear plants where it was made and is currently stored. We should start moving it to Yucca Mountain as soon as possible. Research should continue because more knowledge is good, but the hope that such knowledge will reassure the public is forlorn. To do that will take a president who really understands the issues and knows the numbers.

Depleted Uranium

When uranium-235 is enriched, the uranium-238 is removed. As I mentioned previously, the pure uranium-238 is called *depleted uranium*. Even with the uranium-235 gone, it is still radioactive, with a half life of 4.5 billion years.

Depleted uranium is used in the shells of hydrogen bombs. It also has some important military uses that have nothing to do with its nuclear properties. Depleted uranium is used in the casings of artillery shells designed to attack tanks and other armored vehicles. It is good for this purpose because of two characteristics that help it to penetrate armor. First, it is almost twice as dense as lead. Second, when it hits a metal shield, it tends to form highly concentrated streams instead of spreading out and splattering.

People opposed to using depleted uranium argue that it leaves radioactive material on the battlefield. Proponents say that the danger of radioactivity is small compared to the damage done by war, and that the alternative (lead) is also highly poisonous.

14

CONTROLLED FUSION

IN MANY ways the ultimate nuclear power dream is to harness *fusion*, the reaction in which hydrogen nuclei come together to form helium. This is the reaction that powers the sun. Humans have already achieved fusion on Earth in the hydrogen bomb, but the H-bomb is violent and destructive. Can we also achieve fusion in a controlled way—*controlled thermonuclear fusion* (CTF)? We've controlled fission in nuclear reactors. Why not a fusion reactor? The fuel is abundant; hydrogen is part of ordinary water. Water as a fuel! The idea is wonderful.

Indeed, I believe that nuclear fusion power is in our future. Many of our top scientists even think it is almost imminent—at least something you may see in your lifetime. To make it work, you would first have to chemically extract the hydrogen from the water, and that takes energy, but it doesn't take much energy—at least not compared to the enormous energy you will get back when you burn the hydrogen into helium.

In the hydrogen bomb, the fuel we use consists not of ordinary hydrogen but of the special forms known as deuterium (heavy hydrogen, with an extra neutron in the nucleus) and tritium (doubly heavy hydrogen, with two extra neutrons in the nucleus). These same forms of hydrogen will be the ones used for the first generation of fusion reactors. Deuterium can be extracted from ordinary water; it accounts for about one part in 6700 of the hydrogen. That number may seem small, but it means that each gallon of seawater has enough deuterium to yield the energy of 200 gallons of gasoline. Tritium is not found in water in sufficient quantities, but it can be made—once we have the reactor operating—by bombarding lithium or boron with neutrons. The energy released in fusion is enormous, and the fuel supply on Earth is readily available and essentially unlimited. What are we waiting for?

The fusion reactor has been a dream since the 1950s. The amount of fuel needed is impressively small. For a gigawatt power plant, the total weight of the deuterium and tritium would have to be only 200 pounds per year. We will never run out, at least not in the foreseeable future.

Several technologies are being investigated to make controlled fusion feasible. The main physics problem in fusion is the fact that the hydrogen nuclei repel each other because of their electric charges. In the hydrogen bomb, this repulsion is overcome by the very high kinetic energy obtained from the heat of the primary fission bomb. That's why the hydrogen bomb is called *thermo*nuclear. The high temperature is what makes the fusion work. We can take a similar approach for our reactor, making the hydrogen very, very hot, with a temperature of millions of degrees Fahrenheit. The problem is that any material that hot has high pressure and tends to explode. Moreover, the hot hydrogen will heat, melt, and vaporize any physical container used to hold it.

Three ways of addressing this problem have been proposed. The first is to make the hydrogen gas work at a very low density, so that even though it is hot, the pressure will not get high. This is the *tokamak* approach, named after the Russian device that first achieved some success. The second method is to let the hydrogen explode, but to keep the explosions small. It's analogous to the way we let small gasoline explosions in the cylinders of a car engine run an automobile. That's the laser method, being developed at the Lawrence

Livermore National Laboratory. The third method is even more speculative: to keep the hydrogen cold but use a trick other than heat to eliminate the repulsion of the nuclei. That's called *cold fusion*. We'll discuss each of these methods in turn.

Tokamak

In the tokamak, the hot hydrogen is not solid or liquid but a gas. The gas is so hot that the hydrogen atoms lose their electrons, so technically the gas is a plasma; that is, it is made up of electrons and nuclei that are not bound into atoms. The gas is too hot to be held in an ordinary container, so magnets are used. As long as the nuclei are in motion, the magnetic field applies a force to hold them in. Some people say the hot hydrogen plasma is held in a "magnetic bottle."

Figure 14.1 shows the inside of a tokamak bottle. The person sitting on the left gives an idea of the size, although this tokamak is actually a very small one. The opening is shaped like a doughnut and is kept (when running) at very high vacuum. The hydrogen plasma circles around in the doughnut and is heated by changing the magnetic field. When the plasma gets hot enough, the deuterium and the tritium begin to fuse to make helium and neutrons. Most of the energy is carried off by the neutron and is absorbed by a blanket of lithium. The heat from this lithium is used to generate electricity. The neutron also breaks the lithium nucleus, creating new tritium nuclei to be used as fuel. You can get more than one neutron from each reaction if you let the neutron hit a material such as beryllium, which breaks up and multiplies the number of neutrons, so the tokamak can produce more tritium than it consumes.

Tokamaks are big and expensive, and still very experimental. A very large tokamak called the *ITER* (for *International Thermonuclear Experimental Reactor*) is under construction in France and is due to be operational by about 2016. Figure 14.2 shows a drawing of it. Note the person standing at the lower left. The doughnut-shaped vacuum space has a volume of 30,000 cubic feet.

Figure 14.1. The inside of a tokamak under construction at Princeton. Note the person sitting on the left.

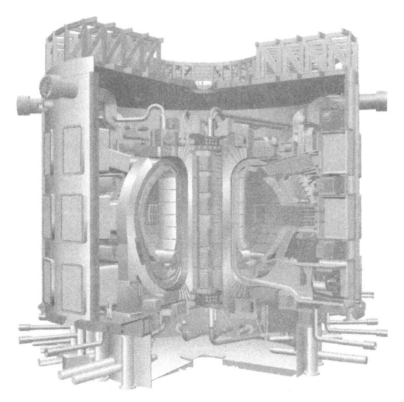

Figure 14.2. The ITER tokamak design. The man at lower left reveals the enormous size of this reactor.

The ITER has the goal of producing a half gigawatt of thermal power for 8 minutes, using ½ gram of DT (deuterium + tritium) fuel. That is a good step toward the eventual design of a power reactor, if it works as hoped. But a true power tokamak is probably at least 20 years in the future, and some people think it will take much longer.

Laser Fusion

Lasers have the ability to deliver large amounts of power to small objects. For this reason, the US Department of Energy has had a major program to see if large lasers (filling a big building) are able to heat a small pellet of deuterium and tritium hot enough to ignite useful thermonuclear fusion. To control the fusion, scientists use small pellets; in other words, they limit the fuel that can ignite in any one explosion and, in that way, control the rate of energy production. Contrast that with a thermonuclear bomb, in which the fission bomb primary has to be a critical mass—that is, big enough to have a chain reaction. There is no good way to contain such a fission primary, so the explosions are always big. With laser fusion, the hydrogen is ignited by lasers rather than by a fission bomb. That has a great advantage in that the lasers don't blow up.

Laser fusion has not yet proven to be a practical method for producing power, although there is reason for optimism. The program at the Lawrence Livermore National Laboratory is now called *NIF*, for *National Ignition Facility*. You can read more about it at www.llnl.gov/nif. The facility fills a building the size of a football stadium with 192 large lasers. Figure 14.3 shows a photo of the building.

The lasers deliver a power of 500 trillion watts (5×10^{14}—that is, 5 followed by 14 zeros); that is 1000 times the electric generating power of the United States. But they deliver that power for only 4 nanoseconds. (A nanosecond is a billionth of a second, about the duration of one computer cycle.) The energy released in this short time is 1.8 megajoules, concentrated into a region about a cubic millimeter in size. This kind of controlled nuclear fusion may be used

Figure 14.3. The NIF laser building at Livermore. It is filled with 192 large lasers.

someday for generating electric power, but the NIF can't be cycled fast enough for that purpose. Fusion ignition is scheduled for 2010, but the pulses of fusion will be few and far between.

The energy sounds less impressive if you calculate that 1.8 megajoules is the same as the energy in 1.5 ounces of gasoline. It isn't a lot of energy. The purpose of the NIF is to deliver this modest amount of energy very, very quickly, so that the fusion will take place before the capsule has a chance to cool off by radiation. It is called the National *Ignition* Facility because it will ignite fusion, even though not in a way that is yet useful for power generation. It is a first step. Fusion power of this kind will probably not be available during your presidency, but you may be asked to supply funding for its continued development.

One of the justifications for the NIF is that it allows us to create small thermonuclear explosions, and we can use the radiation from these to test the susceptibility of military equipment to such effects. By using lasers to implode small pellets, we can also investigate the behavior of new materials that might be used in new nuclear weapons. For that reason, some people oppose the NIF; they feel that it violates the spirit of the Nuclear Non-Proliferation Treaty, even if it doesn't violate the letter of the law. Others argue that the NIF actually helps arms control. Without such a facility, we might have to resume nuclear tests in order to make new bomb designs as part of our stockpile stewardship program.

Cold Fusion

Heat was once thought necessary for fusion because the deuterium and tritium nuclei repel each other. In 1957, Luis Alvarez (my mentor) and his colleagues discovered a remarkable way to cancel that repulsion, thereby eliminating the need for high temperature. The key is an elementary particle called a *muon* that is created in the atmosphere by cosmic rays. It has negative charge, and when it slows down it sometimes sticks to a nucleus.[40] When it sticks to a hydrogen (or heavy hydrogen) nucleus, its negative charge cancels the proton's positive charge. This electrically neutral nucleus then feels no repulsion from other hydrogen nuclei. It wanders around through the fluid (just from its thermal motion) until it gets close to another hydrogen nucleus—close enough for the nuclear force to bring the two nuclei together in fusion. The fusion releases huge energy, and usually ejects the muon. The freed neutron is then attracted to another hydrogen nucleus. It neutralizes that one, and the process begins again. The muon is said to be a *catalyst* because it comes back again and again to keep working its magic. When Alvarez discovered this process and finally understood what was happening (after consulting with Edward Teller), he told me that his excitement was indescribable. It seemed that he had found a simple method of supplying humans with all the energy they would need for the indefinite future.

Unfortunately, it turned out that there was no practical way to make use of this muon-catalyzed fusion. The problem is that the muon sometimes sticks to the fused helium, and then it no longer triggers additional fusions; it stops working as a catalyst. This happens before enough energy is released to make another muon, so there is no net energy gain. Scientists are still experimenting with different pressures and temperatures in the hope that muon-catalyzed fusion will some-day work, but I am not optimistic.

The fact that this technique almost worked has given people hope that other approaches toward cold fusion could succeed. In 1989, two chemists—Stanley Pons and Martin Fleischmann—thought they had achieved cold fusion using a palladium catalyst, but their discovery was based on a mistaken interpretation of their data. Other cold fusion has been reported from time to time. Although there is no proof against it (after all, Alvarez did see a kind of cold fusion), most experts are pessimistic. The reason is that there are no other suitable replacements for the muons used by Alvarez. Other chemical processes typically have energy per atom that is a million times too small to allow the nuclei to approach each other.

The whole field is distorted by the fact that anyone who discovers cold fusion will (1) win a Nobel Prize, (2) become a multibillionaire, and (3) be known in history as the person who solved the world's energy needs. As a result, when someone sees something that looks like cold fusion (but isn't), it is so exciting that there is a strong tendency to want to believe that a real discovery has been made, and to keep all the details secret—until they are patented. The problem is that secret discoveries can't readily be checked by other scientists.

Controlled fusion is in the future. Someday it is very likely to be a major source of energy. Right now, it appears to be more than 20 years away—but it has been in that category for several decades.

NUKES: PRESIDENTIAL SUMMARY

BECAUSE nukes are badly misunderstood by the public, you will have a tough job convincing people to support your policies, no matter what they are. Many people think all radioactivity is bad, and they fear any clicks that they hear on a Geiger counter. If you argue that the waste from a nuclear reactor can be safely stored for long enough, many people will regard you as a dupe of the nuclear industry. You may try to present the numbers only to discover that nobody wants to listen.

Despite the public opposition, nuclear power is likely to be an important part of our future energy needs. Somehow you need to convey to the public that their fear has come from ignorance, not from knowledge, and that you know what you are talking about. Small amounts of radioactivity have such small consequences that they could properly be ignored. Radiological weapons are not the threat that many seem to fear.

On the other hand, nuclear warheads are just as frightening as they have been portrayed to be. Yes, a single weapon can destroy a large city. Yes, potential adversaries still possess such weapons. Perhaps (as of 2008) the danger of a total nuclear Armageddon has been reduced; the United States no longer faces a powerful enemy with thousands of warheads aimed at US cities. Still, never in the history of the world has one object, small enough to fit on the head of a rocket, been capable of instantly killing millions of people. Nuclear arms reduction and other steps to remove the danger of nuclear war will be well worth your effort.

No high school student is going to design, let alone build, a nuclear weapon. It is almost as unlikely that a terrorist group will do so. To the extent that there is any danger, it is that the terrorists will obtain a working bomb from someone willing to sell one—if not North Korea, then possibly a disgruntled watchman from an old Soviet storehouse. If terrorists try to build a nuke and all they do is spread radioactive debris, then the result is not a great threat; it is equivalent to a poor dirty bomb, and although it might spread some fear, most people will be surprised by the absence of dead bodies.

Nuclear proliferation is still a worry. The facilities being built in Iran and elsewhere could easily be modified for making weapons. The hard part of enriching uranium is handling the large amounts you have to process to convert the uranium from 0.7% U-235 to reactor-grade 3% U-235. By the time you've done that, the amount of material you have to handle has been reduced by a factor of four, and further enrichment to 80% or 99% U-235 purity is relatively straightforward. No matter what the intentions of Iran are, the capability to make weapons is being developed in that country. The situation in North Korea is equally dangerous, and its resolution really depends on whether that country will fully dismantle its facilities. Inspections are capable of verifying that, if the inspectors are truly given full access.

Nuclear power in the United States, at least from nuclear fission reactors, will likely be important in the coming years. Reactor designs are far safer than they have been in the past, and if that inherent assurance allows the cost of safety measures to be reduced, then nuclear power could become significantly cheaper than power from oil or even natural gas. The public still worries about waste storage—a concern that I feel has been greatly exaggerated. You need to be a great communicator to convince them that this problem has a reasonable and inexpensive solution.

Controlled fusion power (in any incarnation) is not likely to be practical during your terms in office. You probably should keep up a vigorous research program, however, because surprises in this technology can sometimes solve seemingly intractable problems.

IV

SPACE

Being in orbit is like being infatuated—
you are constantly falling, but you aren't getting closer.

Figure 15.1. Earth from space.

NOTHING has made the Earth smaller than our leaving of it. A ballistic missile traveling above the atmosphere needs less than 45 minutes to reach its target, even if it has to go halfway around the world. In fact, the distance between Moscow and New York City is only 17 minutes by satellite. The image of the Earth from space—the lustrous blue sphere dabbed with white clouds—has become an icon for the unity of the world. Space allows us access to otherwise hidden information. We can watch the development of hurricanes far out at sea, see dust blow from China over the Pacific to the United States, and observe nuclear plants being built in North Korea and Iran. Look at the image of Asia at night that is shown in Figure 15.2. The striking contrast between North and South Korea gives us a dramatic sense of their relative economies that we can't get from scholarly commentary alone.

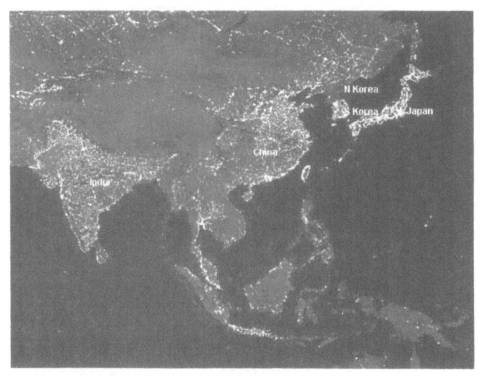

Figure 15.2. Asia at night. Note the difference between North and South Korea.

Space seems limitless, not only in its extent but also in its potential. Yet private industry has been slow to go there. Why? How important is it for us to send humans into space? Should we aim for the Moon? Mars? These are all issues you will be confronting when you are president.

We begin by talking about orbits and satellites—how they work, what they can do, and what they can't do. A full understanding of the physics of a space orbit will be very helpful in understanding satellites, intercontinental ballistic missiles, and spying.

15

SPACE AND SATELLITES

You are in an elevator and the cable suddenly snaps. As you anticipate death, you try to enjoy the fascinating sense of weightlessness as the elevator and you fall together. You float around inside the elevator. There is no force of your weight on your feet. Your shoulders do not feel the weight of your head. That's because your head falls with your chest, with your feet, and with the elevator at the same rate. The muscles in your neck aren't needed to keep your head above your chest. All the time, of course, the Earth is pulling you rapidly toward it. You have weight, but you feel weightless. A movie made of you in the elevator would show you floating around, apparently without weight. Someone watching might think you were an astronaut floating in the International Space Station. In fact, you *feel* just like the astronauts feel when they are weightless. Except, of course, for the dread of impending doom.

Some rides at amusement parks allow you to fall over 100 feet and experience weightlessness, at least for a small number of seconds. I've been on one of those, called *The Edge*, and I confess it was difficult to study the fascinating phenomenon of weightlessness in the midst of the abject terror I felt while falling.

You actually feel the same weightlessness anytime you jump or fall. That sensation is responsible for the giddy feeling you get when you dive into a pool, particularly from a high board. It is amazing how the human body adjusts; after doing many dives, you no longer notice. You can also experience this weightlessness by skydiving or bungee jumping. You do not feel this way when swimming, because your internal organs still press against the outside of your body; your head and your torso are still held up by the force of the water, although a little more uniformly than when they are held up by a mattress. Swimmers are no more "weightless" than you are when you lie on a bed. Similarly, skydivers don't feel weightless once they have reached terminal velocity and are no longer accelerating.

Let's get back to the falling elevator. Now imagine that instead of falling, the elevator is shot out of a gun with you inside and it flies 100 miles before hitting the ground. During that trip, you will again feel weightless, because you are in motion along with the elevator. You and it (and all your body parts) fly in the same arc. Your head and chest are both moving in that arc together; there is no force between them, and your neck muscles can be completely relaxed. Before being sent into orbit, potential astronauts were flown in airplanes following such arcs (the mathematical path is called a *parabola*[1]) in order to see how they responded to the sensation of weightlessness, and to get them accustomed to it.

When you and the elevator are moving together under no other force than gravity (either falling or shot in an arc), there seems to be no gravity. From its absence, you might think you were far out in space, far away from the gravity of any planet, star, or moon. From inside the elevator, you can't tell the difference. Of course, there is gravity; otherwise you would not be falling.

Now imagine that our large horizontal gun is at the top of a very tall tower, 100 miles high. We will shoot the elevator with you inside out of this gun. If we pick a "low" velocity, such as 1 mile per second, you and the elevator will curve toward the Earth and crash into it, as in Path A in Figure 15.3. If instead we pick the higher velocity of 5 miles per second, then you and the elevator will follow

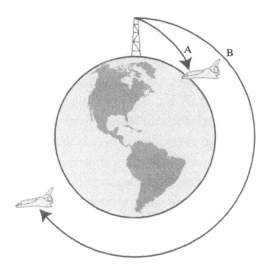

Figure 15.3. The path of a capsule shot from a tower. Path A reflects a lower velocity (1 mile per second); Path B, a higher velocity (5 miles per second).

Path B. You will curve toward the ground, attracted by its gravity, but because of your high velocity you will miss the edge of the Earth. You will keep curving downward, but you will never hit. You are in orbit. The force of gravity makes the path of the elevator—let's call it a space capsule now—curve downward. If that curvature matches the curvature of the Earth, then your trajectory not only misses the surface but stays at a constant height above it. If your velocity is not exactly horizontal, or if your velocity is a little low or high, then the orbit will not be a circle but an egg-shaped path called an *ellipse*.

This may seem preposterous, but it is reasonable to think of an astronaut in orbit around the Earth as being in a state of perpetual falling. That's why the astronaut feels weightless.

You can think of the Moon as doing the same thing—constantly falling. It is attracted to the Earth by gravity, but it has high sideways motion. Even though it is falling toward the Earth, it always misses. This is what Isaac Newton realized in that moment when gravity suddenly made sense to him—when he suddenly understood that the physics of the Moon was the same as that of a falling apple. The Moon is pretty far away—240,000 miles—and gravity is much weaker out there, so the Moon's sideways velocity doesn't have to be so big to keep it up. In fact, it is moving (with respect to us) at only

0.6 mile per second, but because of the weak gravity, that velocity is enough to keep it from crashing down to Earth.

Suppose you are in orbit and you *do* want to come back down to Earth. All you have to do is slow down. Do that and you will no longer be in the orbit (B in Figure 15.3), but in the falling path (A). Some people mistakenly think that the way to reenter is by pointing your rockets upward and firing them toward space to accelerate you downward toward the Earth. Yes, you can do that, but in doing so you will gain velocity. It is better to point your rockets forward and use them to slow you down. Stop your forward motion completely, and you will fall straight down.

In a real rocket, it is more practical to slow down only slightly. That makes your orbit point downward just a little bit, but if you get low enough to run into some of the atmosphere, then the resulting friction will provide the remainder of the slowing. That's how we do it with essentially all of our space satellites. When we want to deorbit, we use retrorockets to reduce the velocity a little, lowering the orbit enough to skim the upper atmosphere, where friction with air provides the rest of the slowing. Of course, it is that friction that poses the great danger to astronauts. Enormous kinetic energy (from the 5-mile-per-second velocity) must be turned into heat.

Although it helps on reentry, atmospheric drag is an impediment that prevents us from orbiting the Earth close to the surface. If the air weren't there, we could skim a mile above the surface, or even a few feet above, with only trees, hills, mountains, and other obstructions causing any problems. Air is thinner as you go higher (it's hard to breathe on Mt. Everest) because gravity is pulling the air down, and it gets even thinner at satellite altitudes. How high do we have to go to get air thin enough to allow us to orbit? At 5 miles per second, the speed required to stay in orbit, even very thin air exerts a big force. As a rule of thumb, it is impractical to orbit lower than about 100 miles.

Orbiting the Moon is different. Because the Moon has virtually no atmosphere, satellites orbiting it can skim the surface, as long as they stay high enough to avoid mountains.

Satellite Orbits: LEO, GEO, and MEO

The velocity required to stay in orbit depends on the altitude because the gravity of the Earth gets weaker at higher altitudes. The Earth is large, so the effect is small until you get very high. At 100 miles altitude, gravity is 95% as strong as it is on the surface. (The reason astronauts feel weightless is not that gravity isn't there, but that they are falling.) At 200 miles altitude, gravity is 90% as strong. That means that the velocity required to orbit is slightly less, by about 5%. Rocket scientists need to know about that decrease, but the important fact for future presidents is that all low Earth orbits (LEOs) require speeds of about 5 miles per second. At that speed, the satellite goes around the 24,000-mile circumference of the Earth in about 90 minutes—1.5 hours. That's why it needn't take more than 45 minutes for a satellite (or nuclear warhead) to go halfway around the globe. The time is about the same for an altitude of 100 miles or 200 miles.

If the satellite moves faster than 5 miles per second, it will leave the circular orbit and head farther out into space. At about 7 miles per second, it will not return but will curve away from the Earth and head into a large orbit around the sun. It will have plenty of velocity to reach the Moon. This higher velocity is called the Earth *escape velocity*. To escape the pull of the sun takes a higher velocity—a solar escape velocity of about 26 miles per second—although some of this can come from the motion of the Earth.

Low Earth orbit is commonly used for spy satellites. The reason is simple: to see best, you need to be close. Telescopes help, of course, but there is a limit to how much an image can be magnified without serious blurring caused by the wave nature of light. The problem with LEO is that you fly past any spot on the ground at about 5 miles per second. Even at an altitude of 100 miles, you can observe that spot for less than a minute. (In 60 seconds you go 300 miles.)

It is possible to hover, but not by coming to a halt. The trick is to use a geostationary Earth orbit (GEO). A GEO satellite is actually orbiting the Earth, but it does this at a high altitude of 22,000 miles, where the gravity is weak,[2] so it orbits at a relatively low velocity (just under 2 miles per second) and takes 24 hours to circle at that greater distance. Because the Earth rotates once in the same period, it will stay above the same location. Both are moving—your home with the TV dish, and the satellite—but their angle with respect to each other does not change.

Weather satellites and TV satellites have this special orbit, and as a result they stay above the same location of the Earth at all times. That's important because it means that the same weather satellite will be able to watch the development of a hurricane, a cold front, or another meteorological phenomenon continuously. If you are receiving a signal for your TV from a GEO satellite, you never have to repoint the antenna. The satellite remains in the same direction above your house at all times.

There is a catch. For the satellite to stay in the same location in the sky relative to the ground, it must orbit above the equator. Can you see why? A geostationary satellite moves in a circle around the center of the Earth. If the satellite is not in an equatorial orbit, then it will spend half of its orbit in the Northern Hemisphere and half in the Southern. Only if it orbits above the equator can it stay precisely above the same Earth location at all times. If you look at them up in the sky, all the geostationary satellites line up in a narrow arc above the equator. That's the only place they can be. That constraint creates a problem. If the satellites are too close together, their radio signals can interfere. So international treaties are required to divide up the space.

Sometimes you will hear these satellites referred to as geosynchronous satellites. NASA uses the term *geosynchronous* to describe any satellite that orbits the Earth in 24 hours, including those in polar orbits. Such a satellite is obviously not geostationary, since it spends as much time over the North Pole as over the South Pole. Only if it is above the equator, is it geostationary.

The problem with geostationary satellites is that they are so far away. The altitude of 22,000 miles is over five times the radius of the

Earth. From way out there, the Earth looks small. That's good, if you are trying to watch weather patterns. One photo can show an entire ocean and still see the eye of a hurricane. Moreover, a weather satellite can be devoted to a special region, such as the East Coast of the United States, or Europe. (But recall, it can't be right above the United States, because it must be above the equator.) Look at weather satellite photos and see if you can tell where the satellite is located. Does the photo look as if it were taken from a satellite right above the equator?

Geostationary satellites have an important advantage for TV broadcasts because to get a signal from one, you never have to repoint your home receiving dish; the satellite always seems to be in the same location in the sky—unlike the sun, Moon, and stars, which appear to move. The disadvantage is that the satellite is 22,000 miles from the Earth, and that means that much greater power is required for its signal to be detectable by your home receiver. For commercial satellite broadcasting, the advantage of constant location in the sky has won out over the power require-ment disadvantage.

Satellite TV is even more popular in developing nations than in the United States. That fact does not reflect physics concerns, but sociology and business; an international company can set up such a system without having to build elaborate infrastructure within a par-ticular country. It can broadcast to nearly half the world at the same time. I recall being really surprised when I visited the ancient and almost unaltered pre-medieval city of Fez in Morocco. The old part of the town has streets and buildings whose appearance has not changed for a thousand years—except, that is, for the satellite dishes recently installed on nearly every rooftop.

Here is an unusual application for geostationary satellites: if you are kidnapped and don't know where you've been taken, try to spot a satellite dish. If the dish is pointing straight up, then you know you are on the equator. If it is pointing horizontally, then you know you are at the North Pole. But be careful; even at the equator the satellite doesn't have to be overhead. The satellite could be above

the Congo, and you could be in Brazil. So you really have to determine the direction of north to make good use of the satellite dish
information.

Between the LEO satellites and the GEO satellites are the MEO
satellites—those in a medium Earth orbit. The most fascinating
ones in this category are the GPS satellites.

GPS: A Set of Medium-Earth-Orbit Satellites

The Seven Wonders of the Ancient World were all architectural,
but I feel that one of the wonders of today's world is the Global Positioning System, abbreviated GPS. A small GPS receiver costs under
$100, and it will tell you your exact position on the Earth within a
few feet. I've used such a receiver in the wilderness of Yosemite, in
the souks of Marrakech, and in the mountains of Rwanda. You can
buy a car with a built-in GPS receiver that will automatically display
a map on your dashboard showing precisely where you are. The military uses GPS systems mounted to its smart bombs to steer them to
hit specific locations within a few feet

A GPS receiver picks up signals from several of the 24 GPS satellites currently orbiting the Earth, whichever ones happen to be in
best view. (Typically half of them will be completely on the other
side of the Earth.) The GPS receiver also contains a small computer.
The system is able to determine its distance to each satellite by measuring the time it takes for a signal to go from the satellite to the
receiver. Once the receiver has measured the distance to three satellites, the computer can then calculate precisely where on Earth it is
located.

To understand how the GPS receiver calculates its position, consider the following puzzle: Suppose I am in a US city. I tell you that
I am 800 miles from New York City, 900 miles from New Orleans,
and 2200 miles from San Francisco. What city am I in? You can use
the map in Figure 15.4 to figure out the answer.

You'll discover that only one location has those distances from

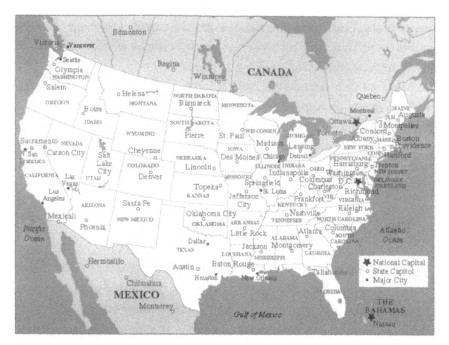

Figure 15.4. Map of the United States. Find the unique location that is 800 miles from New York City, 900 miles from New Orleans, and 2200 miles from San Francisco.

the specified cities: Chicago. The point is that knowing three distances uniquely locates the position. GPS works in a similar manner, but instead of measuring distances to cities, it measures distances to satellites. And even though the satellites are moving, the satellites know and report precisely where they are at any moment. They tell the computer in your GPS receiver where they are so that the GPS receiver can figure out where it is.

You might expect that the GPS satellites would be in geosynchronous orbits. They are not, primarily because such a large distance would require their radio transmitters to have much more power to reach the Earth. They were not put in low Earth orbit (LEO), because if they had been, they would often be hidden from your receiver by the horizon. Each one would be in view for, typically, a minute or two if it passed directly overhead. Instead the GPS satellites were placed in a medium Earth orbit (MEO) about 12,000 miles high. They orbit the Earth every 12 hours. There are 24 such satellites so that your receiver has a good chance of finding three of them at relatively good viewing angles.

Many people mistakenly assume that a GPS receiver broadcasts its own signal to communicate with the satellite. In fact, it is only a receiver, not a transmitter. It could have been designed differently, but the original application for GPS was military, and soldiers don't want to reveal their positions by emitting radio waves.

Spy Satellites

Spy satellites carry telescopes to look down on the surface of the Earth and see what's going on. They were once used exclusively by the military, to see the secrets of adversaries, but now they are widely used by government and industry to look at everything from flooding and fires to the health of food crops.

The ideal spy satellite would stay above the same location all the time, but to do that it would have to be geostationary with an altitude of 22,000 miles. At that distance even the best telescopes can't distinguish small things. Let's digress for a moment to talk about telescope resolution.

The ability of a telescope (or a camera, or even your eye) to resolve objects is limited by the fact that light is a wave and it tends to bend a little when it enters the telescope. That bending blurs the image. To show what's going on here I'll use some equations that future presidents don't need to know, but it will be instructive to see how the calculation is done.

The details of the current US spy satellites are classified, so instead let's work out an example using the Hubble Space Telescope, the multibillion-dollar system used for astronomy. The amount of blurring is given by a simple equation: $b = h \times L/d$, where b is the amount of blur, d is the diameter of the telescope lens or mirror, h is the height of the satellite, and L is the wavelength of light. The only trick to using this equation is that all the units have to be the same.

You can uses inches or feet or fathoms, but you have to choose one and only one unit. I'll choose feet.

The Hubble has a mirror diameter $d = 94$ inches $= 8$ feet (I'll round off for simplicity). We'll assume that we put it in geostationary orbit so that it can continually (at least during the day) spy on Iraq. The altitude is 22,000 miles $= 100$ million feet. (I'll ignore the fact that the place we are observing is not on the equator.) The wavelength of light is $L = 0.0000017$ foot. We plug in these numbers to find the blurring $b = h \times L/d = 100{,}000{,}000 \times 0.0000017/8 = 21$ feet.

That's terrible resolution! The blurring is so bad that two objects closer than 21 feet will be so blurred that you can't tell there are two rather than one. A resolution of 21 feet isn't bad for looking at clouds, watching hurricanes, and maybe even locating ships, but it's not very good for identifying a particular terrorist, or even for counting the number of tents in a training camp.

Now let's work out the same equation for the same telescope but for an altitude of 1,000,000 feet (about 200 miles), a typical LEO spy satellite altitude. Plugging in the numbers as before, we get a resolution of $b = 0.2$ feet $= 2.4$ inches. That's great! You might even be able to identify people. You could, for example, measure a terrorist's height from the length of his shadow.

The important conclusion is that, because of the blurring of light, spy satellites must be in low Earth orbit if they are to have good resolution. Even so, they won't be reading many license plates (despite popular belief that they can). There is no way around the blur problem; it is based on physics.

As I mentioned earlier, spy satellites have a serious problem related to their low Earth orbit: dwell time. Remember, a satellite in low Earth orbit is moving at 5 miles per second. That means it passes over its target very quickly. It can't stop, or it would just fall. Suppose the orbit is 200 miles up. In that case, the satellite may start looking at you 200 miles before it passes overhead (at that distance it sees you at an angle of 45 degrees), and stop looking when it is 200 miles past you. So it examines you while it travels 400 miles. At its speed of 5 miles per second, it does that in just 80 seconds, a little over a minute. If it is twice as high, it will take twice as long to pass through the same angle—160 seconds—but because it is farther away, its resolution is twice as bad, and it will not be able to resolve as much detail.

You can verify this short dwell time yourself. You might be out in the wilderness, watching the sky. A moving star appears. Is it an airplane? No, it really looks like a star. It is likely a satellite. In fact, it is probably a low Earth satellite, since those are the close ones that you can spot most easily. You watch it, but not more than a minute or two, because that is how long it takes to pass at its speed of 5 miles per second.

You have probably had a similar experience looking out of an airplane window. Assume that you are flying at an altitude of 6 miles (about 32,000 feet). You are traveling 600 miles per hour. Then you will go 6 miles in just 1/100 of an hour—that is, in 0.6 minute. You see something interesting right below you, and you tell your friend to look, but if the delay is much more than half a minute, it is too far away to see.

The 1 to 2 minutes available to LEO satellites is a very short time to spy. In fact, many countries that want to hide secret operations from the United States keep track of the positions of our spy satellites (they look up the information on the Internet) and make sure their operations are covered or hidden during the brief times that the spy satellite is close enough to take a photo. Don't get angry when your CIA director can't get the photo you want. It's not his fault, but the fault of physics, and maybe of the hobbyists who measure and post the satellite orbits.

It's even worse than that. If the orbit of the satellite doesn't carry it within about 200 miles of the object, it may be too far away to see clearly, because of the resolution limit of the telescope. Moreover, you can't even take a second look on your next orbit, because the Earth will have spun the object out of the way. It takes the LEO satellite 90 minutes to go around. By that time, the Earth has rotated. The equator goes around its 24,000-mile circumference in 24 hours, so it is moving at 1000 miles per hour. A spot on the equator will move 1500 miles in an hour and a half. It will not be underneath the satellite when the satellite comes back, but 1500 miles away.

This is the dwell time problem. If you have a particular object that you want to view, you may have to change the orbit of the satellite so that it passes over the location. Once it does, it won't pass over that object again (at least not in sunlight) until the object has gone completely around the Earth—that is, 24 hours later—and the satellite has orbited 16 times. If you want to have continuous coverage of a particular location, you must use hundreds of satellites, or give up on spying from orbiting satellites and rely instead on circling airplanes, balloons, or something else that can stay close to one location.

If you wanted to view the entire globe continuously by using lots of satellites, you would need them to be orbiting not much more than 200 miles apart. That would take about 5000 satellites, each as elaborate as the Hubble Space Telescope.[3] Spying from space is not as easy as many people think. It is not true that spy satellites keep a continuous watch over the whole Earth.

The military and intelligence agencies are developing quiet, high-flying drones (airplanes without pilots) to do the most critical kind of spying. These vehicles can be shot down more easily than a satellite, if they are noticed, but they are much cheaper than satellites. In Figure 6.2 I showed an image of a high-flying solar-powered plane called *Helios*. Because such airplanes can dwell forever (at least they won't run out of fuel), they will probably take over a larger and larger part of our spying missions in the future. It also helps that even at an altitude of 65,000 feet, such drones get 16 times closer to their targets than do satellites with 200-mile orbits, so the cameras can be 16 times smaller for the same resolution.

Rocket Science

Rockets are a terrible way to get to space. Typically they waste 96% of their energy. The only reason we use them is that we have no better way to achieve the 5-mile-per-second velocity needed to orbit.

To understand why they are so bad, imagine trying to get into space by pointing a gun downward and repeatedly firing bullets so rapidly that the recoil pushes you upward. Sound ridiculous? Yet that is exactly how rockets work. Rockets fly upward by pushing burned fuel downward. Because the rocket weighs so much more than the fuel that is expelled every second, the velocity gained by the rocket is much less than the fuel velocity—just as the recoil of the gun is less than the velocity of the bullet. As a result, more energy goes into the exhaust than into the payload. That's good for guns (in which the exhaust includes the bullet), but not for rockets. The unhappy result is that the rocket must carry huge amounts of fuel. The mass of fuel used is usually 25 to 50 times larger than the payload put into orbit.

For a long time, this huge fuel-to-payload ratio led people to believe that shooting rockets into space was impossible; after all, how could the fuel be carried if it weighed 25 times as much as the rocket? The problem was solved by using multiple stages, so that the heavy containers that held the initial fuel never had to be accelerated to the final orbital velocity. For example, the space shuttle has a final payload (including orbiter weight) of 68 tons, but the boosters and the fuel weigh 1931 tons—a factor of 28 times larger.[4] Of course, the booster never gets into orbit; only the much smaller shuttle does.

How much better can we do? According to physics, the answer is vastly better. Suppose, for example, that we built a tall building extending all the way to the Moon. From the physics of gravity, we can calculate how much energy would have to be supplied by an elevator: the answer is about 1.5 times the payload weight.[5] That is 30 times less than if we had to use a rocket. But rather than building a tall building,[6] another approach is to hang a cable down from space to haul payloads up. This approach was once called *skyhook*, but it ran into a materials problem: for such a long rope, there was no substance strong enough to support even its own weight, let alone the additional weight of a payload. The discovery of superstrong carbon nanotube fibers (a stringlike version of buckyballs) has revived the idea. But nobody has yet figured out how to make such a fiber that is even an inch long, so this approach is unlikely to be practical in the near future.

A more likely idea for efficient travel to space is to "fly" there on an airplane. Airplanes have two attractive features: they use oxygen from the atmosphere as part of their fuel (so they don't have to carry oxidizer, as do rockets), and they can push against the air instead of having to push against their own exhaust. Of course, wings don't work in space, but airplanes could be used to reach orbital velocities, throw the payload upward, and then have a small rocket change the direction for a circular orbit. Although this approach is possible in principle, the technology to achieve 5 miles per second with airplanes does not yet exist. The concept is being pioneered by the Orbital Sciences Corporation, with the airplane giving the rocket only some of the required velocity, but helping by lifting it above much of the atmosphere. The company uses an L-1011 aircraft to carry and launch a three-stage Pegasus rocket. It has successfully launched over 30 satellites using this airplane/rocket combination approach.

16

GRAVITY APPLICATIONS

GRAVITY is a force, but that wasn't always obvious. Before Isaac Newton's work, gravity was assumed to be simply the natural tendency of objects to move downward. Newton's big discovery was that gravity is a force, and that it is the same force that keeps the Moon in orbit around the Earth. The story that he realized this when he saw an apple falling to the ground may actually be true.

We now know that all objects attract all other objects with a gravity force. As far as we know, it is the only force felt by everything.[7] You probably never notice the gravitational attraction between small things because the force is so weak. If you weigh 150 pounds and are sitting about a yard from a friend of similar weight, the gravitational force of attraction between the two of you is 0.0000001 pound. This may seem small, but such forces can be measured; it is about the same as the weight of a flea. The Earth has a lot more mass than your friend, and for that reason alone it exerts a bigger gravitational force on you. We call this force your *weight*. Think of it this way: the force between you and your friend is approximately the same as the force between a flea and the Earth.

You would weigh less standing on the Moon because the Moon is smaller and therefore doesn't put as big a force on you. If you weighed 150 pounds on the Earth, you would weigh only 25 pounds on the Moon. You haven't changed (you are made up of the same atoms), but the force exerted on you is different. Physicists like to say that your mass hasn't changed, only your weight. Think of mass as the amount of material, and weight as the force of gravity.

Mass is commonly measured in kilograms, abbreviated kg. If you put a kilogram of material on the surface of the Earth, the pull of gravity will be a force of 2.2 pounds. So a good definition of a kilogram is an amount of material that weighs 2.2 pounds when placed on the surface of the Earth. That number is worth remembering, since kilograms are commonly used around the world.[8] Europeans buy their vegetables in *kilos* (the short form for *kilograms*) or in *attos*, which are $\frac{1}{10}$ of a kilogram, about 3.5 ounces. Alternatively, we can specify the mass of an object by its weight on the surface of the Earth—that is, in pounds.[9]

Suppose you weigh 150 pounds on the Earth. Go to the surface of Jupiter, and you will weigh nearly 400 pounds. On the surface of the sun you will weigh about 2 tons—at least for a brief moment before you are fried to a crisp. But in all cases your mass will be unchanged.

The equation that describes the pull of gravity between two objects was discovered by Isaac Newton and is called *Newton's Law of Gravity*. It says that the force of attraction is proportional to the mass. That means that if you double the amount of material, you will double the force. That seems reasonable. Each part of the mass pulls at you, so if there is twice as much, the pull is doubled. Lessen the mass (replace the Earth with the Moon, for example), and the force on your feet (your weight) gets weaker.

The force also depends on the distance. If you are 2 times as far from your friend, the force of mutual attraction is less by a factor of 4. It is now only $\frac{1}{4}$ the weight of a flea. If you are 10 times as far away, the force is reduced by a factor of 100. Note that the reduction is equal to the square of the increased distance. In math we call this an *inverse square law*. The law is inverse because when the distance increases, the force decreases. The law is square because if you triple the distance (that is, increase it by a factor of 3), the force decreases by 9 (that is, 3 squared), and so on.[10]

g's

The force of gravity depends on the mass of the object. A heavier object has a greater force on it. So why do all objects fall at the same speed? The answer is that although heavier objects have a bigger force, they also need a bigger force to accelerate. It takes more force to accelerate a big mass downward than to do that to a small one. The two effects cancel each other, so big and small objects will fall together.

At the surface of the Earth, this acceleration of everything is called the *acceleration of gravity*, given the symbol *g*, and pronounced simply "jee." The actual value of this number is a 22 mile-per-hour increase in velocity every second, written as 22 mph/second. That means that a person falling accelerates from 0 to 66 miles per hour in just 3 seconds. That's why falling is dangerous. For comparison, a Ferrari 333 SP accelerates from 0 to 60 in 3.6 seconds. That's 16.7 mph/second, or ¾ *g*. Of course, the Ferrari does the acceleration horizontally. It is dangerous too.

It turns out to be very useful to think in terms of *g*'s, because that tells you how much force you need to accelerate. When the space shuttle takes off, the maximum acceleration is 3 *g*. That means that a 150-pound astronaut will feel pressed against the seat with a force of 450 pounds, not including the astronaut's weight. That is the maximum that the designers feel the astronauts can tolerate, so that's what sets the acceleration limit.

A fighter pilot sometimes enters a maneuver that will accelerate him (or her, these days), for a brief time, at 9 *g*. If he weighs 180 pounds, the plane exerts a force on him of $9 \times 180 = 1620$ pounds, over ¾ of a ton. Put another way, he will feel, briefly, as if his body weighs ¾ of a ton. Trained fighter pilots can bear this force for a fraction of a second, if they are wearing the right kind of pressurized suit to keep all their blood from flowing to one side of the body.

A world-class sprinter can do the 40-meter dash in 4.4 seconds. His average acceleration is about 9.4 mph/second, about 40% of a *g*.

The following rule is simple to state, but it revolutionized physics. Even if you have taken high school physics, though, you might not recognize it. It is more formally called *Newton's Second Law*.[11] Here it is in the form of an equation: $F = wA$. In this equation, w is the Earth weight (what the object weighs if sitting on the Earth's surface) and A is the number of g's. The equation gives you the force you need to accelerate. We can solve for the acceleration to get $A = F/w$; this equation gives the number of g's you will get from a given force.

The acceleration of 1 g might seem small, but it adds up. Suppose you accelerated at 1 g for a year. How far would you get? The answer is surprising: about ½ light-year![12] That's the distance that light travels in ½ year. The distance to the nearest star (not counting the sun) is about 4 light-years.

Sometimes people make the mistake of thinking that an object in space can be accelerated very easily because it is "weightless." But that isn't true. (If it were, it would be very easy to accelerate to the high speeds needed to reach distant stars.) The force needed to accelerate an object is exactly the same whether the object is on the surface of the Earth or in space. To accelerate an object to 1 g, we have to apply a force equal to the weight that it would have on the Earth's surface, even though the object is in space.

Using Gravity to Search for Ice, Oil, Hidden Craters, and Secret Tunnels

I said earlier that every object exerts a small gravitational force on every other object. Remarkably, measurement of such small forces has important practical applications. If you are standing over an oil field, you will feel slightly less gravity than the gravity you feel standing over solid rock. That's because oil is less dense than rock, so it has less mass than if the space were filled with rock. Such small gravity changes can be measured using a *gravimeter*, an instrument that measures the pull of gravity on a weight. A map of gravity over the surface of the Earth can reveal the density of the material under the ground. It is, in some ways, a great way to "X-ray" the Earth, because the effects of gravity pass right through the ground. Geologists commonly use gravimeters to search for likely oil locations.

The strength of gravity can even be measured from space. A pair of satellites called GRACE (*Gravity Recovery and Climate Experiment*) were orbited to measure the gravity pattern over the whole world. These satellites recently found that the ice volume in Antarctica is decreasing by 36 cubic miles per year! That amazing result is important in the discussion of global warming, a topic we'll return to in the last section of this book. Given the remoteness of the region and the complexity of the ice flow, it is hard to imagine any other technique that could have been used to deduce such a key number in such a clear way.

The GRACE satellites are so sensitive that they can even detect the change in gravity over Brazil from the extra water that's there after the rainy season. Take a look at the maps in Figure 16.1. The shading represents the excess water in the region, measured from its gravity. Compare March and August in the middle of the continent.

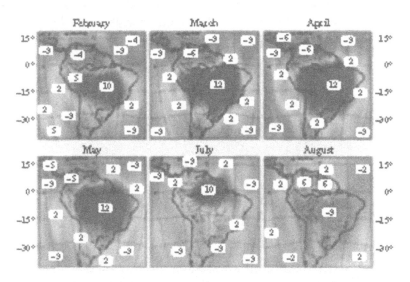

Figure 16.1. Water in the Amazon region, detected by the GRACE satellites from its gravitational force. The difference between March and August amounts to about an inch of water on the surface.

The difference in water averages about an inch. Here we have another wonder: one inch of water change detected by its gravity effect in space! (Of course, it is one inch spread out over a very large area.) You can even see this water draining down the Amazon, particularly in the July image.

In a more surprising use of such gravity measurements, a map was made of the buried Chicxulub crater on the Yucatán Peninsula—the crater left behind when an asteroid killed the dinosaurs. The crater was filled in by sedimentary rock that was lighter than the original rock, so even though it is filled, it shows a gravity anomaly—that is, a difference from the gravity force expected if the rock were uniform. An airplane flying back and forth over this region made sensitive measurements of the strength of gravity; the results are shown in the computer-generated map shown in Figure 16.2.

In this map, the high regions are those in which the gravity is stronger than average, and the low regions are locations where the gravity is slightly weaker. The crater shows several concentric circles, the largest over 60 miles in diameter. The inner rings probably

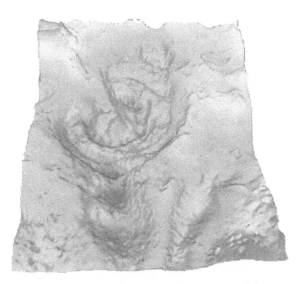

Figure 16.2. A gravity map showing the rings of the buried Chicxulub crater. This is the crater formed by an asteroid impact 65 million years ago, resulting in the demise of the dinosaurs.

formed when material under the huge crater was forced upward, partially filling it.

In an underground tunnel the dirt that once filled it is not there, so the force of gravity on the surface is reduced compared to nearby regions. It is a small effect, but it is measurable, and it has practical importance for future presidents. Gravimeters driven on trucks have detected such weakened gravity and discovered clandestine tunnels under the border between the United States and Mexico—tunnels used for smuggling drugs and people. Using other techniques, the United States has also found clandestine tunnels under the demilitarized zone separating North Korea from South Korea. Some of these tunnels were large enough to allow passage of a tank, and North Korea apparently built them to use in the event of war with South Korea. Needless to say, we would like to know if there are any other undiscovered tunnels of this kind, and scientists are attempting to use gravity as a way of detecting them.

Space Industry: Manufacturing in a Weightless Environment

When the space program began, many people thought there would be significant advantages to being in a weightless environment. In a satellite, things wouldn't sag under their own weight, so it might be possible to make better (rounder) ball bearings, or to grow more perfect crystals for use in computers and other electronics.

This promise has been largely unfulfilled. The additional cost of doing the work in a satellite has not turned out to be worth it. It costs about $5,000 to launch a pound of anything into orbit. In the near future, commercial companies hope to reduce that price to $500 per pound, but even so, it's hard to make a profitable factory in space when it costs so much just to get there. In principle, there is no reason why getting to space must be expensive. To achieve orbit, the space shuttle uses fuel weighing 28 times as much as the payload, but in principle, for an efficient system the fuel need weigh only 1.5 times as much. Sometime in the future, if travel to space becomes as cheap as an airplane ride, the idea of factories in orbit may be revived.

Gravity on the Moon and Asteroids

Our Moon has about $\frac{1}{81}$ the mass of the Earth, so you might think its gravity would be 81 times weaker. But the Moon's radius is 3.7 times smaller than that of the Earth. When these two effects are combined,[13] the result is a surface gravity $\frac{1}{6}$ that on the Earth.

With weak gravity, you might expect people on the Moon to be able to move faster than those on Earth. But it is tricky to walk in low gravity, until you get used to the fact that a person who weighs 150 pounds on the Earth weighs only 25 pounds on the Moon. The astronauts who landed on the Moon in 1969 seemed to move in slow motion. When they jumped they went high (for people in bulky space suits), but when they came back down they came down slowly. They found it easier to hop than to walk.

Imagine now an asteroid with a radius of 1 mile. How heavy would a 150-pound person be on this asteroid? We can use the gravity equations to calculate the value,[14] and the answer is interesting: about half an ounce. That's the weight of five pennies on the Earth. You would have to be very careful on the asteroid. Your escape velocity would be very low, and you could easily launch yourself into space by jumping. To jump into space from the Earth, you would require a velocity of 2700 miles per hour (assuming no air resistance); but from the asteroid, only 6 miles per hour—about the speed of a person jogging. This low escape velocity was a problem for a US space probe called the *Near Earth Asteroid Rendezvous* (NEAR). If the satellite had hit its target asteroid Eros with a velocity of 6 miles per hour or more, then it might have bounced right back out into space.

The space shuttle weighs 2 tons. If you normally (on Earth) weigh 150 pounds and you are 10 feet from the center of the craft, then the shuttle will attract you with a force of about $\frac{1}{10}$ of a gram. That's about the weight of three staples. There is a force—just not a very large one.

One of the most common conceits in science fiction movies is the implicit assumption that all planets in all solar systems have a gravity about equal to that of the Earth. There is no reason why that should be so. Pick a random planet, and you are just as likely to be a factor of six times lighter (and bouncing around like astronauts on the Moon) or six times heavier and unable to move because of your limited strength. Imagine a person who weighs 150 pounds on the Earth trying to move on a planet where he weighs 900 pounds.

Artificial Gravity

Even when the actual gravity is negligible, one can achieve a similar effect by accelerating the spacecraft. Science fiction movies often show "gravity" in spaceships. Some scientists claim that these movies violate laws of physics, but these scientists are wrong; they are making the implicit assumption that the spacecraft is not accelerating. It is reasonable to assume that future ships are powered by antimatter and have lots of energy available. A little of this energy could easily be devoted to keeping the ship in constant acceleration, by an amount of 22 miles per hour every second, or 1 g. In this scenario the astronauts would feel the force of the floor beneath them as the ship accelerated them, and it would be indistinguishable to them from gravity. This may be an option if we ever make long journeys into space. Remember what we calculated earlier: if you accelerate for 1 g for 1 year, you go half as far as light would go in that same time.

This acceleration may also be achieved in a sideways direction. Remarkably, this is possible without the addition of energy, because accelerating to the side can put you in circular motion without changing your speed. This trick was portrayed accurately in the 1968 movie 2001: A Space Odyssey. If you are on the outer part of a spinning ship, the walls put a force on you to keep you in circular motion. The same thing happens when you are in an automobile that is making a sharp circular turn. You feel the force of the car against your side as it pushes you to travel in the circle with it.

Physicists sometimes like to analyze problems in rotating coordinate systems. When they take this approach, they introduce a concept called *centrifugal force*. Although many high school teachers will tell you that centrifugal force doesn't exist, that's only because they don't want you to attempt calculations in rotating frames—a task considered too advanced. But if you are in a rotating spaceship, then the rotating frame is very convenient, and from that point of view, the "gravity" you experience is just the centrifugal force.

We can also think about Earth satellites by using the concept of centrifugal force. Imagine that you have a rock tied at the end of a string and you are spinning it in a circle above your head. The string provides the force that keeps the rock from flying away, and this force also keeps the rock in circular motion. If the string breaks or is released by the slinger, the rock flies off in a straight line. Such a sling was the weapon that, according to the Bible, David used to kill the giant Goliath.

Gravity does the same thing for an Earth satellite: it provides the force that keeps the satellite in a circular orbit. When gravity balances the centrifugal force, the distance from the Earth to the satellite remains constant.

In a similar manner, if we could suddenly turn off the force of gravity, the Moon would leave its circular orbit and head off in a straight line. The same is true for all the satellites now in orbit around the Earth. And with the sun's gravity turned off, the Earth would head out into space too, at its previous orbital speed of 20 miles per second.

The X Prize

How high do you have to go to qualify as an astronaut? The answer depends on your definition. You never get far enough to completely escape the Earth's gravity, and likewise, no matter how high you get, there will always be some atmosphere (even if it's only one molecule per cubic mile). Because a satellite orbiting at an altitude of 60 miles will stay up for at least a few orbits before being brought down by air resistance, some people have defined this altitude as the beginning of *space*. The United States is a bit more generous. It awards "astronaut wings" to those who travel above an altitude of only 50 miles.

To encourage private companies to develop rockets or aircraft capable of reaching space, a private foundation created the X Prize for the first commercial organization to send a human up 100 kilometers—that is, about 60 miles. I won't go into all the details of this prize because it is far less interesting than most people think. There is a huge difference between reaching a 60-mile altitude and orbiting the Earth. Way back in 1946, the US government launched a modified German V-2 rocket to an altitude of more than 100 miles, but it was not until 1957—11 years later—that the Soviet Union launched the first Earth-orbiting satellite. The United States followed with its first successful orbit in January 1958. Getting into orbit is much harder than merely reaching space.

To get a feel for the relative difficulty of actually going into orbit, imagine that there is no atmosphere on the Earth. Suppose we want to shoot a bullet straight up and have it reach an altitude of 60 miles. How fast must it leave the gun? The answer is about 1 mile per second.[15] Compare this to the speed needed for orbit: 5 miles per second. Moreover, the kinetic energy depends on the square of the velocity, so the energy needed is a factor of 25 times bigger. Thus, it takes 25 times as much energy to get into orbit as to merely reach a 60-mile altitude. That means that the rocket system that won the X Prize had only 1/25 of the energy needed to get its "astronauts" into orbit.

Why is there so much excitement over the X Prize? In part it is because very few people know about this factor of 25. I've found that even many professors are surprised. The winners of the X Prize certainly understand, but they don't want to publicize that. The $10 million prize was not enough to cover their expenses. How will they make money? Not by sending up satellites. They can't accomplish that any more than the United States could with the V-2 rocket. The interesting answer is that the winners of the X Prize will earn money by launching paying customers to an altitude of 60 miles, and then awarding them astronaut wings. They make their money on tourism—just like the fictional investors planned to make their money in the novel *Jurassic Park*.

A future president, however, must know that getting into orbit takes enormously more energy than just getting into space. Small rockets, such as those that won the X Prize, are extremely far from being able to orbit satellites.

17

HUMANS IN SPACE

ON February 1, 2003, the space shuttle *Columbia* exploded upon reentry over Texas, killing all seven astronauts. This tragic loss stunned the nation. Why? Not why did the shuttle explode, but rather why did its loss stun the nation? It shouldn't have. The dangers of shuttle flight are obvious. The tragedy should not have been a surprise to anyone, certainly not to a future president.

Here is the physics: at 5 miles per second, every ounce of the shuttle, including its human cargo, carries over 10 times the energy of an equal weight of TNT.[16] When the shuttle is launched, that energy is in the fuel; when in orbit, the energy is stored in its motion. The launch delivers the fuel energy to the shuttle, and reentry takes it away. If any of that transfer goes awry and the energy is dumped into heating the shuttle, then tragedy is certain. That's what happened on January 28, 1986, during the launch of *Challenger*, and on February 1, 2003, during the reentry of *Columbia* (Figure 17.1). Just as with an airplane, takeoff and landing are the most dangerous phases of flight.

If the energy is turned into heat and the heat is transferred to the shuttle, then its temperature will rise to over 140,000°F, more than 10 times hotter than the surface of the sun[17]—hot enough to vapor-

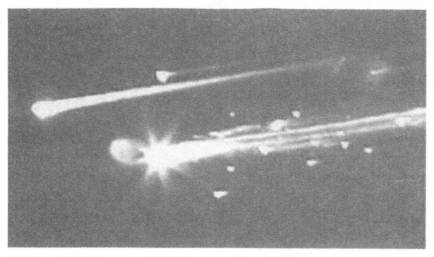

Figure 17.1. The explosion of the space shuttle Columbia *on February 1, 2003.*

ize every known material. The shuttle survives reentry by making sure that most of the energy goes into heating the atmosphere, not the ship. Hobbyists who photograph safe reentries typically see a beautiful bright trail made of ions created from the intense heat of air impacting the shuttle. The shuttle looks and behaves like a large meteor. Astronauts peering out their windows have been equally entranced (and maybe a bit frightened) as the dark sky of space is hidden by the glowing hot plasma. Can such a reentry be made safe? Yes, someday, but we are not close to that day yet. To me, the remarkable achievement of the manned space program is that only two shuttles have been lost in more than 100 missions.

Few jobs are more dangerous than that of an astronaut,[18] but most of the public doesn't know that—or rather, didn't. The misimpression that the shuttle is safe is largely the work of NASA public relations, which has strived to make the flights sound routine. The shuttle is so safe, they suggest, that we can fly senators and school-teachers. Shuttle flights are so inexpensive and routine, they want us to believe, that we can use them to perform experiments suggested by high school students.

Another major misconception, which NASA has spread even

more actively, perhaps because many in the agency have convinced themselves that it is really true, is that the primary goal of the space shuttle is the advancement of scientific knowledge. Most scientists roll their eyes when they hear this claim. Don't misinterpret what I say; good, even great, science has been launched into space on the shuttle. Results from the Hubble Space Telescope excite physicists and astronomers as much as the general public. Moreover, the Hubble telescope required repair by shuttle astronauts. Doesn't that justify men in space?

Not really. The military occasionally loses spy telescopes, and its response is to launch a replacement. Launching two completely new Hubble telescopes—the original and a replacement, with neither qualified for human servicing—and therefore cheaper—would arguably have been less expensive in the long run than launching and repairing one. But if NASA had done that, then it could not have used the Hubble Space Telescope as part of its justification for launching astronauts.

When it comes to the science itself, the space shuttle is a poor choice of platform. Humans are a source of noise; they cause vibrations and emit heat radiation (infrared), and even their flea-sized gravitational field changes as they move around, limiting the sensitivity of some instruments. Just flying an experiment on a manned mission automatically raises the experiment's costs because of the extreme safety requirements. Many scientists moan privately about scientific missions that were delayed and made more costly because they were moved off unmanned launch vehicles and forced to become part of NASA's scientific justification for the shuttle. I vividly remember how the COBE (Cosmic Background Explorer) mission, awarded a Nobel Prize in 2006, was delayed because of the NASA-perceived need to launch it with the shuttle instead of a cheaper unmanned rocket.

It is said that there is no substitute for the flexibility and observational abilities of a man in space. That is partially true, but sending people into space comes at great cost, both human and financial. A

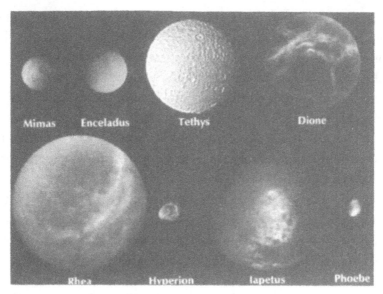

Figure 17.2. Some of Saturn's moons. The biggest surprise was that they are all different from each other.

similar argument was made about horseless carriages: horses are smart, and automobiles are dumb. Most factories now use robots instead of people; just watch the Discovery Channel series *How It's Made* to see what I mean. Until recently, most astronomers felt compelled to spend their nights at the observatory. Now they program the computer and let the robotic telescopes do the job. If science is your goal, launch robots, not people.

Hubble aside, what would you name as the really glorious achievements of the space program in the last 20 years? My favorite is the discovery that every moon of every planet is significantly different from every other moon (see, for example, Figure 17.2)—a result completely unanticipated and still not fully understood. One might also pick the amazing success of weather satellites. Or the remarkable pictures you get from your satellite TV system. Those in the know might pick our space spy systems. Then there's GPS, the Global Positioning System, used to guide airplanes, boats, hikers, and automobiles, as well as soldiers and smart weapons. Perhaps you would pick the exploration of Mars, with our rovers moving about sampling and analyzing the Martian soil and atmosphere. These projects have one thing in common: they were all unmanned.

After the *Columbia* catastrophe, Senator John McCain said, "Space exploration is a mission the U.S. will not abandon." I hope he is right. But (and here Senator McCain might differ), such exploration in the next decade or two could almost certainly be done better with unmanned vehicles. We are already planning to send robots to Mars to bring back samples. Someday we may even send astronauts there. Let's not be in a hurry to send the humans. New telescopes and unmanned instruments will tell us more about space than orbiting astronauts can. There is a future for humans in space. Eventually hypersonic flight will be perfected, and we will be able to ride airplanes into orbit without having the enormous fuel loads that rockets require. Such planes will have to be able to fly at 18 times the speed of sound, Mach 18, 5 miles per second. A technology to do that, called a *scramjet*, is being developed, but nobody knows when it will be deployable. Scramjets are still in the research and development stage, with strong emphasis on *research*. Tests reveal critical new physics with each added Mach number, and unmanned prototypes have already reached Mach 10. When scramjets become available, human visits to space may become both cheap and safe.

I have spoken to many people in NASA who believe that the future of their agency depends on continuing and extending human space flight. Without astronauts, the argument goes, the public will lose interest in space. I think this argument is wrong. Few Americans even knew that the space shuttle *Columbia* was in orbit in February 2003—until the astronauts were killed. When the students I know at Berkeley put space posters on their walls, they are not posters of astronauts but images of regions where stars are being born, of exploding stars, of extremely distant and hauntingly beautiful fields of galaxies. These images were all taken with instruments that didn't need humans to get into space, and benefited from the absence of humans in order to take stable pictures.

The space shuttle is not safe, and it will not be safe in the foreseeable future. Is using it worth the lives lost? We must be honest. Perhaps the shuttle missions should continue, but if they do, let's do it with full public realization that the chance of death on each mission is about 2%. Soldiers going into battle often accept risks even higher than that. The astronauts who have died in space travel always knew this danger, and they chose to accept it. Can the public accept such a high level? I don't know. But whatever the decision is, it should be made in candor and in truth. The space shuttle is big engineering; it is the dream of man in space; it is an adventure. But it is not safe, it cannot be made safe, and it is not done for science. It is important that future presidents know this.

18

SPYING WITH INVISIBLE LIGHT

You might think that spying is done mostly by looking, but that's not so—at least not if you mean seeing with visible light. Targets of interest, including ships, broadcast stations, military facilities, and nuclear plants, emit some of their most interesting signals in colors invisible to the human eye. These include infrared (heat radiation—indicating signs of life and activity), radio (carrying secret communications), radar (revealing where something is and what it is measuring) and X-rays (which can look right through metal). All of these signals are actually forms of light, but with frequencies (the technical term for *color*) too low or too high for the human eye to sense. I call these radiations *invisible light*. Sensitive instruments can pick them up, measure and map them, and even use them to produce photographs—photographs taken with invisible light.

Let me begin with an anecdote. In 1989 I had an opportunity to spend a night with the US Border Patrol that guards the boundary between the United States and Mexico near San Ysidro, California. After touring their facilities and having dinner, we went to a hillside overlooking the border just as the sun was setting. Many people were gathering on the Mexican side. There were stands selling tacos and hot dogs, primarily (I was told) to people who had been caught the night before and had spent a day waiting for nightfall so that they could try again.

It began to grow dark, and the Mexican side of the border was getting crowded. I could still see everyone clearly. Suddenly one boy ran to the fence, climbed over, and ran to hide on the US side. That seemed to trigger an avalanche. Hundreds of people swarmed toward the fence—young and old, some with ladders—and in a few minutes they were all across and disappearing into the gullies of the desert on the US side.

My hosts, the border patrol, did nothing for a while, and then they drove me along a dirt road up to a hilltop. By the time we got there, it was dark. The lights of Tijuana twinkled in the distance, but the desert between us and the border was dark. On the back of a jeep, the border patrol mounted a special pair of night vision binoculars and started using them to survey the darkness. The binoculars were cooled with liquid nitrogen. I was allowed to use them myself to scan the countryside. Through the binoculars I could see mostly blackness, the outline of the hills, and, in the valleys (from our high vantage point), clusters of people glowing in the dark. Their faces and hands were bright, and the rest of their bodies, covered with clothing, were somewhat dimmer. They were waiting. At one location they had lit a small fire (visible even with the unaided eye as a small reddish spot), which was an extremely intense white in the binoculars.

"What are they waiting for?" I asked.

"Their guides," the border patrolman replied. The immigrants had been given simple maps to show them how to get away from the fence, to a location that they could find less than a quarter mile in from the border. There they would be met by the guide they had hired.

They waited a long time, and so did we. Finally, after over an hour, the guides had found the groups, and together they started moving through the gullies. I wondered if they knew how easily we could see them in the absolute darkness. As one group approached a road, we drove over to them. They heard the vehicle coming and waited.

"Why don't they run?" I asked.

"It's too dangerous," the border patrolman answered. "They might get lost, and besides, if they do get caught, we don't put them in jail. We only send them back to Mexico. Then they can try to get across again tomorrow."

(A physics book is not the right place to discuss the wisdom of US border policy, so I include an additional observation in the notes.[19])

Figure 18.1 shows an image of people taken with a night vision telescope—not sneaking into the United States, but climbing a fence. It was staged to show what such an image looks like; the telescope was similar to the one I used on the border, and it does remind me of what I saw that night.

Figure 18.1. Night vision: an infrared image of people cutting a fence and climbing over. Note that the warm parts of the image (face and hands) are brighter than the cooler outside surface of clothing.

Infrared Radiation

The mysterious binoculars that I used at the border were an optical system that could see infrared (IR) light, also called *heat radiation*.

Such light has a lower frequency than that of visible light. Humans emit it because we are warm. The electrons in our atoms shake from their heat, unless they are at absolute zero. Shaking electrons have shaking electric fields, and shaking electric fields create electromagnetic waves. The effect is analogous to the emission of all other waves: shake the ground enough, and you get an earthquake wave; shake water, and you get a water wave; shake the air, and you get sound; shake the end of a rope, and a wave will travel along it; shake an electron, and you get light. Shake it gently, at low frequency, and you get IR light. You can do that with low heat, such as the heat of the human body. At higher temperatures, the electrons shake faster, and that creates higher-frequency light. Heat something over a flame, and it gets red hot. Higher temperatures yield yellow light, then blue, then blue-white. Color—related to frequency—tells you the temperature. That's useful.

Humans emit about 50 watts, the equivalent of a small lightbulb. That seems like a lot, and it is, but the number shouldn't surprise you. You know that the presence of a person can significantly warm a small room. Moreover, you know that an adult, even if inactive, eats about 2000 food calories per day. That is equivalent to a total energy use of about 100 watts,[20] out of which half goes into heat radiation. Of course, our bodies are usually surrounded by objects (for example, clothing, a house, the ground) that are only a little cooler than we are, and we absorb radiation from them. Yet this number illustrates why humans can barely survive on 1000 food calories per day. That much is radiated away every day, so we need more food in order to do work. In the United States, an active teenager consumes about 3000 food calories per day, 50% more than an adult.

Because 50 watts of IR is a lot, it is not difficult to devise an instrument to detect it. Once it is measured, the signal can be used to drive a TV and make an image that I can see. That's what I was actually looking at through the binoculars that night near the Mexican border: a small TV screen. In late 2001, when the United States went after Osama bin Laden in the Tora Bora mountains of Afghanistan, newspapers reported that he would be hard to find in the snow-covered mountains. In fact, warm people are particularly easy to spot when everything else is cool, because their infrared radiation stands out in stark contrast. Even caves containing people emit strong signals from their openings—stronger than empty caves. Of course, the military did not catch him, but it wasn't because he was hard to see. The problem was determining which one was him.

Remote Sensing of the Earth

Infrared satellites can measure the temperature of the Earth's surface by the amount of infrared radiation emitted. Figure 18.2 shows satellite measurements of the temperature of the ocean surface, called *sea surface temperature*. High sea surface temperature in the Caribbean during 2005 was detected in this manner and led scientists to predict a dangerous hurricane season. That prediction, unfortunately, came true.

Figure 18.2. The Earth's surface temperature pattern, measured from a satellite with an IR camera. The warmer regions emit more IR, and so appear brighter in the image.

Modern weather satellites take images in visible light and in the infrared. The visible light is primarily reflected sunlight (it also includes electric lights and fires), and it shows such things as land, water, and clouds. Infrared comes from the thermal IR radiation of the atmosphere and ground, and for that reason it can be used to measure the temperature. (Some of it also comes from reflected sunlight.) Weather satellites also include a kind of infrared, multispectral sensor that is sensitive to special wavelengths that give additional information about the surface. One of the most important of these is a camera that is sensitive to IR light that has a frequency about 12 times lower than that of visible light (wavelength of 6.5 microns)—a frequency that is emitted by water vapor. Air that is high in water vapor will form clouds (and rain) if it is cooled, so a photo showing the flow of water vapor, like the one in Figure 18.3, can be used to help predict where storms might form.

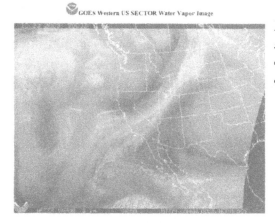

Figure 18.3. Weather satellite image showing the water vapor path over the western United States.

Military Special Ops

The US Special Operations Command has the motto "We own the night."[21] There is physics behind this motto: they can see at night by using night vision viewers worn over their eyes. Owning the night is more than just having these viewers, though; it is being trained to use them comfortably and effectively. Special Ops uses two types of IR viewers.[22] One is sensitive to low frequencies,[23] and picks up the infrared radiation emitted by warm people. This was the kind of viewer that I used at the Mexican border. The other type picks up higher-frequency IR, sometimes called *near IR* because its frequency is near that of visible light.[24] Special Ops teams carry flashlights that emit near IR light; it is invisible to enemy eyes but can be seen through special infrared viewers. I own a video camera that has a near IR lamp built in. It is advertised as being able to take movies in total darkness, and it can, provided that *total darkness* means there is no light visible to ordinary human eyes. Near IR flashlights are also used to provide the signal for TV remote controls. Click a button on your remote, and a series of IR flashes is emitted. Your receiver detects these flashes, recognizes the pattern, and changes the channel.

The lower-frequency *far IR* viewers are valuable for detecting anything that is warmer than the surroundings. An infrared viewer, looking at an automobile, can tell if that automobile has been operated recently because it shows the warm region near the engine. An infrared viewer can detect the heat from a campfire long after it has been extinguished. Search on YouTube and you'll find lots of movies taken in the far IR. Systems like this are used by the police to detect unusual power activity in a house. People with illegal greenhouses in their home (to grow some illegal plants) have been caught because the roofs of parts of their houses showed unusual heat radiation. Similarly, IR viewers can locate human activity in remote jungles and other places.

The United States operates unmanned aircraft called *drones* or *UAVs* (for *unmanned air vehicles*) that carry infrared imaging devices. The most impressive of these is the Global Hawk drone, shown in Figure 18.4.

During the war in Afghanistan, a Global Hawk UAV flew from Germany to Afghanistan, directed only by GPS satellites, and then circled over Afghanistan for 24 hours and flew back. It is even capable of landing at its airport itself, unguided by humans. While above its targets, its cameras (infrared and visible) and radar constantly sent images back to the United States by radio. It flew at an altitude of 65,000 feet, the same as the U-2 spy plane, and far above the abil-

Figure 18.4. The Global Hawk unmanned air vehicle (UAV).

ity of the Taliban to shoot down. In fact, it flies so high that they probably never even knew when it was watching them. From the angle of the tails visible in the photo, the drone appears to be incorporating stealth technology to make its radar signature weak; we'll talk about such design in a moment. Stealth design is also the reason the engine is mounted on the back of the plane, where it will not be easily visible to ground radar.

Stinger Missiles

One of the real worries of the American troops in Afghanistan is the threat of Stinger missiles. They can be launched from a device held by a single soldier and aimed at a low-flying airplane or helicopter. A Stinger missile weighs only 35 pounds and can reach up to about 10,000 feet. These missiles were given to the Taliban by the United States back in the days when the United States was trying to undermine the Russian-imposed government. A device mounted on the Stinger missile steers the missile toward anything up in the sky that is emitting strong infrared radiation. That means anything hot, such as the tailpipe of an airplane or helicopter. To prevent a Stinger from hitting his airplane, the pilot sometimes drops hot flares so that the missile will choose to go after those instead. Some advanced Stinger missiles have special devices to prevent them from being fooled in this way. Because the Stinger is so small, it can change direction more rapidly than an airplane or helicopter, so it is hard to outmaneuver.

Stinger missiles have become so widespread (70,000 have been produced) that there is a real danger that they will be used by terrorists against commercial aircraft in the United States. Their range is limited to about 3 miles, so the terrorist would probably position himself just outside the airport and try to hit the airplane just after it takes off. Over its history, the Stinger missile has been used to shoot down over 250 aircraft. It is possible (but not known for sure) that a Stinger missile was used in 1994 to shoot down the airplane carrying Rwandan President Juvénal Habyarimana just before it landed at the Kigali airport, with that assassination serving as the trigger for the horrible genocide that followed.

Humans aren't the only animals that use IR. Pit vipers and mosquitoes also detect the warmth of IR by special sensors, and they use this signal to help find their prey. Human lips are also sensitive, and we use them to judge the temperature of food when it is close to our mouths. Try the following experiment: close your eyes, and put your lips close to those of another person. Notice how easily you can sense the warmth.

Radar and Stealth

The word *radar* was originally an acronym for *radio detection and ranging*. It is a method of emitting radio waves (primarily microwaves) into the sky and looking for reflections off metal objects such as airplanes. From the time it takes the signal to come back, the distance to the plane can be measured. Microwaves are used for radar because of their short wavelength, typically about one inch. That means that they don't spread too much from a medium-sized antenna (several feet across), so you can aim them accurately and determine the direction of the plane. For airplane detection it is also valuable that they travel relatively easily through clouds and smoke.

Radar was invented and used for the first time during World War II to detect Nazi airplanes crossing the English Channel. It saved Britain from invasion in the following way: During the early parts of the Battle of Britain, the Nazis sent numerous bombers to attack London and other cities. Every time they reached the English shore, however, they were met by British fighter airplanes. The Nazis assumed, incorrectly, that Britain had thousands of such airplanes because they seemed to be everywhere. In fact, there were not very many planes, but radar told the planes where to go to meet the incoming attackers. As a result of this deception, the Nazis overestimated the British military strength and postponed an invasion that otherwise might have succeeded.

Radar bounces off almost everything solid or liquid, so it returns signals from birds and rain, but it tends to bend around objects (such as fog droplets) that are much smaller than the wavelength of the microwaves. All large airplanes are equipped with radar to detect other airplanes and stormy regions. You can buy a complete radar system for a few thousand dollars for your pleasure boat, and use it to navigate through fog and spot other boats before they get too close.

When radar bounces off a moving object, the returned frequency is changed slightly. When the object is moving toward you, each subsequent cycle of the reflected radar signal doesn't have as far to go as the prior one, so the pulses arrive closer together. That gives the signal a higher frequency. If the object is moving away, the reflected signal is lower. This change in frequency is called the *Doppler shift*, and radar that takes advantage of this is Doppler radar. Such radar is used by the military to measure the velocity of airplanes, and by the police to measure the velocity of your car. Measurement of the Doppler shift from raindrops in a hurricane allows the wind velocity from a satellite to be measured.

Stealth is an important technology that has been under extensive development by the US military for several decades. It is a method to make airplanes, ships, and other objects "invisible" to radar. You may be surprised to learn that the primary secret of stealth is to cover the vehicle you wish to hide with mirrors.

Mirrors? Don't they reflect signals? Yes, but unless they are facing you, they deflect the signal away. It is easy to hide in a hall of mirrors. Stealth takes advantage of the fact that essentially all radar and infrared detection systems are *monostatic*, meaning that the transmitter and receiver are at the same location. In radar, the same dish is used to send out the beam and to concentrate the returned signal bounced off the target.

On a normal airplane or boat, some part of its curved surface is likely to send some of the received signal back at the radar transmitter. That is the signal that the radar detector uses to tell that you are there. The returned signal is even stronger if there are right-angle corners on your vehicle, since they have the property that incoming waves are reflected right back at the source, after bouncing off all three walls. Corners like this are called *retroreflectors*, and most airplanes and ships have some corners somewhere on their surface. Figure 18.5 shows a diagram of a corner reflector.

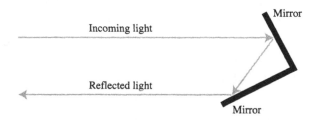

Figure 18.5. Light bouncing off a right-angle corner.
It returns in the opposite direction from which it came.

Stealth vehicles are different. They are constructed with *no* right-angle corners and mostly flat surfaces. Take a look at the Nighthawk fighter shown in Figure 18.6. Note the flat surfaces. Each one reflects radar, but except for that brief moment when the surface is completely flat toward the source, the signal does not return to the

Figure 18.6. Nighthawk stealth fighter. Its surface is covered with flat radar mirrors which, ironically, make it immune to radar.

sender but bounces off in another direction. Note the splay of the tail wings—designed so that they do not make right angles with each other or with the body of the plane.

Look back at Figure 18.4—the Global Hawk drone. Note that the tail is quite different from the configuration in commercial airplanes. The usual arrangement contains two right angles, formed by a vertical structure called the stabilizer and two small horizontal wings with flaps called elevators. In contrast, the Global Hawk and the Nighthawk have no right angles.

There are other stealth tricks too, such as using materials that do not reflect as strongly as metal does. But the key idea in stealth is to be careful with the geometry. For the Nighthawk shown in Figure 18.6, the real trick was to make a flat-surface airplane that was still capable of flight as a fighter. Newer stealth airplanes (such as Global Hawk) have cut back on the flat-surface requirement. Even a curved surface, if properly done, will not reflect a signal back to the source.

Ships at sea are also detected primarily by their radar reflection. A stealth ship designed by the flat-surface principle is shown in Figure 18.7.

Figure 18.7. A stealth ship. As with the Nighthawk fighter, the flat surfaces reflect radar, but usually not back to the transmitter.

Radar Imaging

When visible light reflects from an object, a camera can use that light to form an image—a photo. An IR telescope forms an image with infrared. The same can be done with radar. In fact, radar is now being used extensively as an imaging tool and for mapping. It is possible to use radar from an airplane to form an image of the ground that includes hills, buildings, even cars and trucks. Since radar penetrates clouds, such images can even be made on a foggy day. To make it even better, the image can be accumulated as the airplane flies, and the total signal (stretched across 5 minutes of flight) can be put together to form the same kind of image that a huge radar receiver would have obtained—that is, the image that would be produced by a radar receiver several miles long (as long as the flight path). Such a system is called *synthetic aperture radar* (SAR) because it synthesizes the image from many reflected signals measured as the aircraft flies near the target. SAR systems have also flown in satellites, including the space shuttle.

Figure 18.8 shows two SAR images: one of the Pentagon (taken by a Global Hawk unmanned air vehicle), and one of New York City. Look at the huge amount of information in those radar images. Would you have guessed they were taken with microwaves? Radar can also penetrate a thin layer of leaves—enabling you to see hiding tanks and artillery in a forest. Radar is no longer just a blip on a round screen, but a highly advanced technology that can be used for surveillance and spying.

A

Figure 18.8. Radar images taken using synthetic aperture radar (SAR).
(A) The Pentagon in Washington DC. (B) New York City.

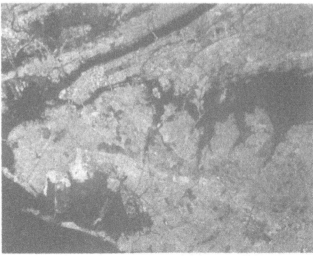

B

X-Ray Backscatter

As I discussed in the section on nukes, X-rays are a form of high-energy light particles; because our eyes can't detect them, they, too, fall in the category of "invisible light." X-rays pass through light elements such as the hydrogen, oxygen, and carbon that make up our flesh, but they are absorbed by heavy elements such as the calcium in our bones. When X-rays pass through matter, the shadow that results is called an X-ray image. Such images have multiple uses for medicine, industry, and national security.

Some very important applications come from the fact that not *all* of the X-rays pass through the light elements; some will bounce off the electrons that orbit the atoms. Send a beam of X-rays toward a person, and a few will bounce straight back to you—and any detector that is nearby. The more electrons the matter contains, the more backscatter there will be. As a result, the amount of backscatter roughly indicates density.

Intelligence agencies can use X-ray backscatter to look inside things even if they can't get behind them to get a shadow. X-ray backscatter can be used to read the combination of a safe, for example, by looking into the workings and seeing how all the elements are arranged. In fact, high-security safes all have a layer of lead in front of the tumblers to absorb X-rays before they can backscatter off critical components.

X-ray backscatter has also been used on trucks to look for smuggled contraband and more. The spectacular image in Figure 18.9 revealed illegal immigrants attempting to enter southern Mexico from Guatemala in a truckload of bananas. This was not an image that could be taken in the United States, since our laws prohibit the use of X-rays without permission, even for non-citizens.

It is obvious that this is a backscatter image because we cannot see very far through the truck; the reflection is off the first foot or two of material.

Figure 18.9. X-ray backscatter image revealing illegal immigrants attempting to enter southern Mexico from Guatemala in a truckload of bananas.

X-ray backscatter can also be used to search a person for concealed weapons, particularly in countries where it is legal to irradiate people without their knowledge. A physicist friend told me he was asked to hand his passport over a tall barrier when passing through a checkpoint at a Chinese airport. He saw no valid reason for the peculiar layout, other than the likelihood that a backscatter X-ray machine had been installed behind the wall and they wanted to check him for concealed weapons with his hands raised; from his on-the-spot technical evaluation, that scenario certainly would have fit.

SPACE: PRESIDENTIAL SUMMARY

RECENT presidents have had a lot of trouble deciding what to do about space. They think of John F. Kennedy's 1961 proposal that the United States put a man on the Moon and return him safely by the end of the decade, and the galvanizing and energizing effect of that idea on the mood of the United States.

Kennedy's idea had a long list of virtues. What he proposed was certainly doable, and not at an extraordinary cost. The technology required already existed, and all it would take was putting everything together and making it work as a package. The project had an aspect of competition with our Cold War enemy, the Soviet Union, and it had a time deadline. It also had a very clear goal. In the mid 1960s, the public felt it was a project that could redeem our country.

Redeem? Yes, I use that word because I believe the enthusiasm for the space program came largely from our wish to put a man on the Moon to honor the assassinated Kennedy, to fulfill his dream, even though he wasn't there to do it himself. Indeed, once the goal pronounced by Kennedy had been reached, the public quickly lost interest in having humans on the Moon (we haven't had any since 1972), or even in space.

Since that era, NASA and other government entities have extolled space for its scientific use. Yet if you ask virtually any scientist associated with the space program, the bigger driving force seems to be to find an excuse to put men back in orbit, possibly in the forlorn hope that the excitement of the Apollo program will be revived and the US public will again be enthusiastic about a big space program.

I don't think that is very likely. Space is an excellent platform for what it does well: spying, weather imagery of the Earth, measurements of the atmosphere, global positioning, satellite communications around the world (including TV and radio), detection of signals from outer space, and all sorts of other scientific endeavors. All of these are best done with robotics.

If I may offer some unsolicited advice, I would recommend that, when you are president, you lower the priority of putting humans into space. Make science truly the primary goals of the government science program. Whenever anyone suggests that a task requires a human, scrutinize it with great care to see if the additional cost really is worth it or should be spent in other ways, such as to fund duplicate or redundant satellites. My sense (poorly developed, I know, since I am not a politician) is that the public thirsts after science. They love the space images, the discoveries, the unmanned robots exploring mars. Their only real interest in astronauts is their safety; can they be brought back home safely? When you become president, don't bring our astronauts back safely; rather keep them here safely, and let their work be done by our magnificent human-designed, human-engineered, human-operated robots and computers.

V

GLOBAL WARMING

PERHAPS *the most difficult issue to face a future president will be global warming. Everybody who is concerned about world events already knows a lot about this subject. But, as our most recent Nobel Peace Prize laureate, former Vice President Al Gore, says in his powerful Academy Award–winning documentary,* An Inconvenient Truth *(2006), the trouble with most folks isn't so much their ignorance, but in knowing so many things that "ain't so."[1] Nearly all who talk or write on the subject put their own spin on it. As a result, even if you are well-read on global warming you probably have accumulated a great deal of misinformation. As a future president, you can't afford to confuse the authentic with the apocryphal. To lead the public down a difficult but necessary path, you need to know the whole truth.*

Here are the key facts that drive our concern about climate change. The average temperature of the Earth is now the warmest that it has been in 400 years. It has increased by about 2°F in the last century. That doesn't seem like a lot, and in some sense it isn't. The fear is that a substantial part of the increase is likely to have been caused by human activity, particularly by the burning of fossil fuels. If that is truly the cause, then we expect the temperature to keep rising. In the next 50 years, the best estimates are that the increase will be between 3°F and 10°F. That is a lot. Already, warming in Alaska from 1900 to the present has been enough to cause significant portions of the permafrost to melt. A 10°F rise would be enough to make fertile regions in the United States arid and trigger large-scale economic disruption around the world. There is also good reason to believe that the warming will be more intense in the polar regions.

Moreover, people have reported a large number of anomalous weather conditions. In An Inconvenient Truth, Vice President Gore showed increases in the intensity of hurricanes, tornadoes, and wildfires. Much of what he says is exaggerated; I'll discuss the details in the chapter titled "A Very Likely Cause." When such exaggerations are exposed, some people are tempted to dismiss the danger altogether, but that is false logic. Incorrect reasons put forth to substantiate a hypothesis do not prove the hypothesis false. There is plenty of reason for concern, and a future president must be willing to act. Of course, the actions must be driven by an understanding of what is real and what isn't. Some proposed actions are merely symbolic; others are designed to set an example; others have the purpose of being a first step. Few of the proposals (and virtually none of those presently being put forth by major politicians) will really solve the problem. As a future president, you need to know the difference between symbolic gestures and effective action.

To make matters worse, the burning of fossil fuels has another effect beyond global warming—one that gets attention in scientific circles but is not widely appreciated by the public. About half of the carbon dioxide emitted from fossil fuels winds up in the oceans, and that makes the oceans more acidic. The problem is not as immediate as acid rain, but still it can affect life in the oceans in potentially disastrous ways. The acidification of the oceans may be a bigger danger to the ecosphere than a few degrees of additional warmth.

Every few years, a prestigious international committee makes a new assessment of the status of climate change: what we know about it, what the consequences are likely to be, and what we can do. This organization is commissioned by the United Nations and the World Meteorological

Organization, and is called the Intergovernmental Panel on Climate Change, or just the IPCC. The IPCC attempts to do the impossible: reach a consensus among hundreds of scientists, diplomats, and politicians. As a result, its conclusions are often muted and mixed, but its reports contain a wealth of data that help everyone evaluate what is going on. The IPCC shared the 2007 Nobel Peace Prize with Al Gore. You need to know the initials. Memorize them: IPCC. You can't discuss climate change at a serious level without knowing about the IPCC, any more than you can discuss world events without knowing about the UN. I'll talk a lot about the IPCC reports in the following chapters.

Weather is highly variable, and so is climate. To understand the most immediate dangers, the long-term ones, and the possible actions we can take, it is essential to distinguish true human-caused phenomena from natural fluctuations. I begin with a brief study of what we know about the climate of the past.

19

A Brief History of Climate

As I mentioned in the introduction to this section on global warming, it is warmer now than it has been for at least 400 years. This is the key fact of global warming and has been recently endorsed in a review by the US National Academy of Sciences, the most prestigious group of scientists in the United States. I accept their conclusion, but not because there is a consensus on this issue; consensus has been a notoriously poor guide for truth, as Al Gore vividly shows in his movie.[2] Rather, I base my acceptance on my own carefully considered judgment of the data and the analysis. I am knowledgeable in this subject; I was the coauthor on a technical book about past climate, and I served as a reviewer for the academy report.[3]

The best data come from thermometer records covering the period from 1850 to the present; written records of temperatures in the early 1800s have not yet been analyzed. A proper analysis is not trivial to do. The Northern and Southern Hemispheres don't show exactly the same behavior, probably because two-thirds of the land mass is north of the equator. And care must be taken not to give too much emphasis to cities. Cities are often referred to as *heat islands* because human materials such as asphalt on the streets absorbs more sunlight than the flora they replaced, so cities are hotter than the surrounding countryside. Hot cities are more of a local effect than a sign of global warming.

Figure 19.1 shows the thermometer temperature based on the 2007 report of the IPCC. The bold line is a running average, to guide your eye so that you can see the trend.

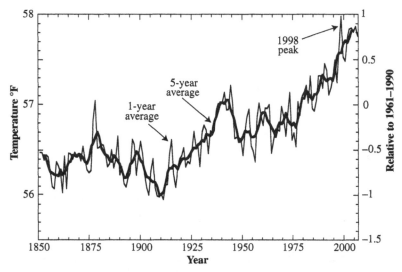

Figure 19.1. Global warming seen with thermometer measurements.

The thermometer data reveal several very interesting things. The temperature from 1860 to 1910 (left side of the plot) was about 2°F cooler than it is now. That number represents an average; some areas did not warm up as much (for example, the contiguous United States), and others warmed more. The coolness of the previous centuries in Europe was enough to repeatedly freeze the Thames in England, and to ice over the canals of Holland during most of the winter. Without such cold we wouldn't have stories like *Hans Brinker, or, the Silver Skates*. The canals rarely freeze over these days. The chill was the lingering end of the "Little Ice Age," a cold spell that took place all around the world. The length of this period is disputed. In *An Inconvenient Truth*, Al Gore shows a plot with the cold weather extending back 1000 years, and he makes fun of those who claim there was a medieval warm period, but the plot he shows has now been discredited, in the National Academy review that I mentioned earlier. I'll discuss that update later in this book, in the section titled "The Hockey Stick." Prior to the Little Ice Age, there *was* a medieval warm period, about as warm as today, that took place from about AD 1000 to the 1300s. According to the National Academy review, there is so much uncertainty in temperature when we go back that far that we can't really tell whether this warm spell was worldwide. I don't think the issue is very important for future presidents. In fact, the last 10 years have been the warmest in four centuries; that we can say. Whether they are the warmest in a millennium is not really critical, except perhaps because it provides a catchier sound bite.

The medieval warm period was certainly present in Europe. There is lots of evidence in indirect records—including the widths of tree rings, a study of light versus heavy water in glaciers, and historical documents—showing the duration that northern harbors were free of ice. It was a time of great productivity in Europe, and some historians refer to it as the *medieval optimum*—a term that annoys those who are trying to warn of global warming. Historian Barbara Tuchman describes how the end of this warm period led to the misery of the Little Ice Age:

> A physical chill settled on the 14th century at its very start, initiating the miseries to come. The Baltic Sea froze over twice, in 1303 and 1306–7; years followed of unseasonable cold, storms and rains, and a rise in the level of the Caspian Sea. Contemporaries could not know it was the onset of what has since been recognized as the Little Ice Age, caused by an advance of polar and alpine glaciers and lasting until about 1700.[4]

Of course, the ice age was not *caused* by the advancing glaciers; rather, the advance of the glaciers was a symptom of the cooling. Cause and effect are frequently confused, and future presidents need to keep them straight.

Some people think that global warming is not caused by human activity, but that the Earth is simply still recovering from whatever natural phenomenon caused the Little Ice Age. The IPCC can't rule out that possibility but gives it only a 10% chance of being true.[5] If the rise is natural, then we are lucky; if past records are an indication of the limits to the natural variability, then the rise in temperature is unlikely to continue much further. In its latest study, however, the IPCC found that it is 90% likely that humans are responsible for at least some of the observed global warming of the last 50 years. Even though it will be expensive to act, a 90% chance is something that a president can't ignore.

Take another look at the thermometer data in Figure 19.1, but this time concentrate on the most recent 50 years, the rightmost part of the plot—the part that shows (according to the IPCC) very likely human-caused warming. Notice that the warmest year on record was 1998.[6] It may seem odd that, with all the global warming taking place, the warmest year was actually in the last century, not this one. But that is not a proper concern. The temperature change is not smooth but bumpy, with peaks and dips that depart from the average. We don't know what causes such fluctuations. The source may be natural variability in cloud cover. If you flip a coin 100 times, you don't always get 50 heads and 50 tails. Likewise, if the climate is changing, some years will still be warmer and some cooler than the trend. Figure 19.1 shows that the natural variations fluctuate typically 0.2°F to 0.4°F away from the curve that represents the average.

Look at the temperature plot again, but this time concentrate on the period 1940–1980. What you see will serve both as a lesson and as a cautionary tale. Around 1942 the temperature reached a high and then began to drop precipitously. In 10 years the temperature dropped by 0.7°F. Back then, geologists and oceanographers were beginning to pull out of rock and sediment the detailed record of past ice ages. What they saw startled and shocked them. Ice ages were not rare events but had been the norm for the past million years. In fact, they appeared to last about 90,000 years, on average, and then to be broken by a brief warm period lasting only 10,000 years—brief respites that they called interglacials. Then the cycle repeated, every 100,000 years.

Here's the part that scared people the most in the 1950s: The last ice age ended about 12,000 years ago. Since then, we have enjoyed about 12,000 years of relative warmth (the Little Ice Age was *very* little compared to the real ice age). But the average duration of prior interglacials suggested that our time of warmth should have ended several thousand years ago. They wondered, in the 1950s, had the warm period finally ended? Was the drop in temperature the beginning of a new ice age?

The cyclic pattern of ice ages had previously repeated itself 10 times, over the past million years. An old theory by the Serbian geophysicist Milutin Milanković was invoked to explain the cycle. It turns out that the orbit of the Earth is not constant but has a wobble (due to the gravitational pulls of Venus and Jupiter) with a quasi-100,000-year period. It seemed that the ice ages were recurring because changes in the orbit of the Earth—something totally beyond our control—had altered the amount of sunlight hitting the Northern Hemisphere.

If you study human prehistory, the danger looked even scarier. Agriculture had been invented, or discovered, just at the beginning of the current warm period, 12,000 years ago. All of civilization was based on agriculture because efficient production of food is what allows a minority to provide for the sustenance of the majority, and that means there will be food for merchants, artists, and even physics professors. And now, in the 1950s, the dropping temperature made people wonder if the ice age was coming back. It wouldn't be a little ice age, but a far bigger one, with temperatures dropping by 10°F, or maybe 15°F.

I was in elementary school at the time, and one of our textbooks had a drawing of the consequences to New York City, with 1000-foot glaciers toppling skyscrapers. Figure 19.2 shows a similar image that appeared on the cover of *Amazing Stories* magazine. It shows the

Figure 19.2. A cover of Amazing Stories *magazine from the 1950s: the ice age returns to New York City.*

Woolworth Building in New York being toppled by a returning gla-cier. This image doesn't seem so fanciful if you walk around New York City with the eye of a geologist.

Figure 19.3 is a photo that I took in the Bronx in 1956, when I was 12 years old, showing deep scratches left in the rock by a prior glacier. Look at the grooves coming toward you from the north, carved in the rocks in the center of the photo. All over the city, the scratches pointed in the same direction, emphasizing the enormous size of the glacier—far taller than the Empire State Building would later be, as it scraped the surface clean. This glacier also left behind

Figure 19.3. A photograph of St. Mary's Park in the South Bronx, taken by the author from his bedroom window in 1956. The grooves visible on the rocks were made by a mile-high glacier that scraped over the region 12,000 years ago.

a pile of debris now known as Long Island, and a series of puddles called the Great Lakes.

In the 1950s, newspapers promulgated another explanation suggested by some scientists: that the cooling climate came from nuclear bomb tests polluting the atmosphere. The United States and the Soviet Union ended atmospheric testing in 1963; France continued until 1974, and China ended in 1990. Linus Pauling won the Nobel Peace Prize for his role in bring about this cessation.

To the relief of many people, temperatures began to rise again after 1970. The ice age was not imminent. Even though the cooling ended, no scientist today believes that the nuclear tests were at fault.[7] Correlation does not imply causality. Many experts now attribute the brief cooling spell to an unusual number of volcanic eruptions that took place during those decades and spewed dust high into the atmosphere. Such material tends to reflect sunlight and thereby reduce the insolation—the solar energy reaching the ground. Once the dust settled and the volcanic activity ceased, the Earth began to warm again.

The rise in temperature continued, and now we are worried about warming. Is this a continuation of the prior trend, the finale to the Little Ice Age? Or is it the beginning of something more ominous? Our now deeper understanding of climate leads most scientists today to believe the latter. In the next chapter we'll discuss how the burning of fossil fuels could be the cause of global warming. It is wise, however, to retain some humility, and to recognize that even a theory that explains what is happening may not be correct.

The Climate of the Past 14,000 Years

To look back at truly ancient climates, we can't rely on written records, but we can extract meaningful measures from old ice. Dig down into the glaciers covering Greenland and you can find ice that was formed 500, 1000, 14,000, even 80,000 years ago. For the younger ice, you can tell its age by counting layers; the yearly cycles of snowfall and dust leave clear marks, analogous to tree rings. The climate is recorded in trapped air, and in the very makeup of the ice itself. Figure 19.4 shows the result: the deduced temperature change over the last 14,000 years.

Study this plot as you studied the earlier one—the one that went back only to 1850. The plot begins on its left side with a dramatic cold period, from 12,000 BC to just after 10,000 BC. That was the end of the last real ice age (not a little one). The temperature was about 12°F colder than the temperature today. So much seawater was bound up in glaciers on the land that the oceans were over 300 feet lower than they are today—enough to allow *Homo sapiens* to walk across the Bering Strait from Asia to America. Barely visible on the right side of the plot is the little dip that we call the Little Ice Age. The big ice age makes it look insignificant. Even so, remember that that little dip caused the "miseries" in Barbara Tuchman's quote. It is hard to conceive what miseries we would have to endure

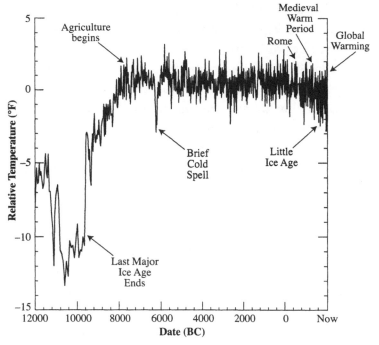

Figure 19.4. The Earth's climate for the past 14,000 years, deduced from Greenland ice.

if a big ice age returned—or if we experienced a similar amount of warming.

Just after 10,000 BC, the suddenness of the rise shows that the ice age ended very abruptly. We think this change was triggered by a change in the Earth's orbit. We don't know why it occurred so rapidly, but most people guess that when land becomes exposed without its icy cover, more sunlight is absorbed on the bare rock than on the prior ice. That accelerates the warming and melting even more. This is an example of *positive feedback*, something that amplifies and accelerates change. A similarly rapid melt is observed in the rivers and streams of Canada every year. As spring begins, the ice on rivers endures until it suddenly cracks, with a bang sometimes heard for miles. Within a few days, the ice is gone and the river flows freely. Positive feedback.

Perhaps the most interesting part of Figure 19.4 is the period covering the history of civilization, the period from 8000 BC, about when agriculture began, to now. The temperature was pretty constant, at least by prior standards. On this plot, global warming of the past 100 years looks quite small. It may now be warmer than it has been for the past 400 years, but temperature throughout the history of civilization has undergone enormous fluctuations, very likely greater than we are now experiencing. Look at the dip near 6000 BC. That appears in many other records too, and no one knows what triggered it. It was very short-lived, possibly lasting only a century or two.

As small as it is, the Little Ice Age does appear to be anomalous. Global warming is quite small. But don't get complacent. If the warming continues in the manner that the models predict, then the expected temperature rise will be 3°F to 10°F. That amount is comparable to the rise at the end of the last ice age. Although that rise may have made civilization possible, it is very likely that a similar rise now would badly disrupt civilization as we know it today.

20

THE GREENHOUSE EFFECT

WALK into a real glass-covered greenhouse and feel the stunning warmth and humidity. Enter an automobile parked in the sun and feel the oppressive, even dangerous heat. Now walk outdoors on a spring day and feel the pleasant temperature of the earth. In all three of these cases, the warmth is a result of the *greenhouse effect*—the same greenhouse effect that scientists now blame for the current global warming, the same greenhouse effect that will (unfortunately) be one of the biggest issues you will have to handle during your presidency.

The greenhouse effect is real and indisputable. It happens whenever energy gets in more easily than it can get out. Think of the parked car. Sunlight streams in through the windows. Some is reflected back out, but most of it is converted into heat—heat of the seats, the steering wheel, and the air inside. Hot air rises, so if you crack open the window a bit, it escapes and cool air flows in. A tiny opening can make a big difference. To cool a car quickly, open the sunroof. To cool a house, open an upper window.

Similar physics warms the Earth. Sunlight heats the surface of the Earth and the air above it. There is no glass to keep the air in, but gravity serves the same purpose, so the heat can't get out by convection. There is only the vacuum of space outside, so the heat can't conduct away. In fact, the only way heat can get out is by IR— infrared heat radiation. The Earth emits IR, but air absorbs it before it can get to space; unlike sunlight, air is opaque (black) to IR. The absorption of IR from the Earth warms the air even more, and the increased air temperature in turn warms the surface. This is the *blanket effect*—called the *greenhouse effect* when caused by sunlight. Energy is reflected back, giving us more heat—as illustrated in Figure 20.1.

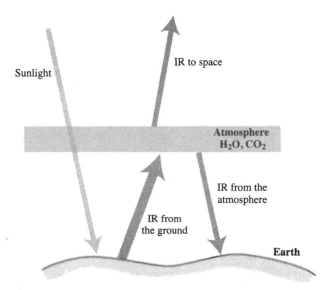

Figure 20.1. The physics of the greenhouse effect. Sunlight passes right through the atmosphere and warms the Earth, but the IR radiation emitted by the Earth is absorbed by air, and some is reflected back down. As a result, the blanket of air keeps the Earth's surface warmer than it would otherwise be.

The Earth's atmosphere is 99% nitrogen and oxygen. Remarkably, neither of these two gases absorbs IR, so they don't contribute to the greenhouse effect. The absorption is all done by trace gases, primarily water vapor, carbon dioxide, methane, and ozone, as well as some others. These gases are known, collectively, as the *greenhouse gases.* To the extent that these gases are a natural part of the atmosphere, we have a natural greenhouse effect. In fact, if not for these gases, the surface of the Earth would have an average temperature of 12°F, 20 degrees below freezing! Look again at Figure 20.1. The surface of the Earth is receiving heat not only from the sun, but from IR emitted by the atmosphere.

The greenhouse effect is one of the fundamental facts of atmospheric science. It is real; that fact is beyond dispute. Without it, the entire surface of the ocean would be frozen solid. Life—at least the kind that depends on liquid water and warmth—could not survive. We owe our existence to the greenhouse effect.

So why are we worried about it?

The answer is that some of the heat radiation leaks out through the atmosphere, because there is not enough water vapor, carbon dioxide, and other gases to absorb all of the IR. Think of the atmosphere as a leaky blanket. This more accurate picture is shown in Figure 20.2.

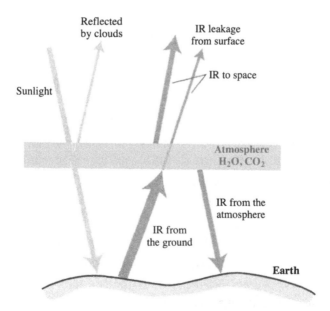

Figure 20.2. The physics of the greenhouse effect, with cloud reflection and atmospheric leakage included.

This figure shows two additional subtle but important effects. Not all of the sunlight reaches the surface; some is reflected by clouds. In addition, not all of the IR emitted by the Earth is absorbed by the atmosphere; some leaks through directly to space. Increase the clouds, and it will get cooler.[8] Plug the IR leaks, and the Earth will get warmer.

We are currently doing just that, plugging the leak—not on purpose, but inadvertently. We are making the atmosphere into a better blanket—by pumping in carbon dioxide and other greenhouse gases. That's the reason we are worried about the greenhouse effect. Remember, the basic greenhouse effect is real, responsible for the comfortable warmth of a spring day and the possibility of life on Earth. If we make the greenhouse effect stronger, the surface temperature of the Earth will rise. The IPCC estimates that the current rate of carbon dioxide injection will do a good job of plugging the leaking IR, and that will cause a rise in temperature somewhere between 3°F and 10°F during your lifetime.

Carbon Dioxide

Carbon dioxide is created whenever carbon is burned. As its name suggests, a molecule of carbon dioxide consists of one atom of carbon and two (that's the *di-*) of oxygen, giving it the chemical symbol CO_2. Burn carbon, and you release both energy and CO_2. We can separate the carbon dioxide back into its components, but only by putting back in the energy we took out. If we have used the energy—for example, to make electricity—we are stuck with the CO_2.

Carbon dioxide is a tiny constituent of the atmosphere—only 0.038%—but it is enormously important for life. This trace gas is the primary source of our sustenance. Virtually all of the carbon in plants, the source of our food, comes from this tiny amount in the air. Plants use energy from sunlight to combine CO_2 with water to manufacture hydrocarbons such as sugar and starch, in a process

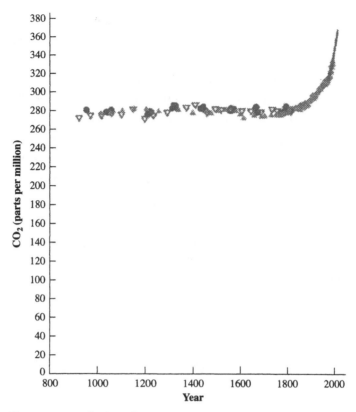

Figure 20.3. Carbon dioxide in the atmosphere over the past 1200 years. The sudden 36% rise in the recent past is due primarily to the burning of fossil fuels.

called photosynthesis. These hydrocarbons are the building blocks of our food and fuel. Photosynthesis also releases oxygen into the atmosphere. When we breathe in oxygen and combine it with food, we get back the energy that the plants absorbed from sunlight.

Scientists traditionally refer to 0.038% as 380 parts per million, abbreviated 380 ppm. Figure 20.3 shows how this level has changed over the past millennium. The amount of carbon dioxide was pretty constant from AD 800 until the late 1800s, at a level of 280 ppm. In the last century it has shot up to 380 ppm—an increase of 36%.[9] If we continue to burn fossil fuels, we expect the carbon dioxide to keep rising.

It's the recent rise that concerns people. Other measurements (not shown) tell us that the carbon dioxide level now is higher than it has been at any time in the last 20 million years. That fact is not disputed; it is astonishing but not surprising. The carbon dioxide comes from human activity, including the burning of fossil fuels and the destruction of enormous regions of forest, primarily in South America and Africa. The latter cannot continue long, even if not stopped by conservationists, because we will run out of forests. In contrast, we will not run out of fossil fuels—at least not coal—for centuries. If we do nothing to stop it, the increase in carbon dioxide is expected to continue.

Until 2006, the United States was the biggest source of the carbon dioxide increase, contributing about 25% of the yearly additions. In 2006, China surpassed the United States, and its contribution continues to grow. China is building the equivalent of 50 to 70 new gigawatt (very large) coal-burning plants every year. Just one gigawatt coal plant burns a ton of coal every 10 seconds.[10] Add in two oxygens from the atmosphere to make CO_2, and that means 3 tons of carbon dioxide every 10 seconds, for *each* plant. World total power production is about 1000 gigawatts.

This carbon dioxide is being dumped into the atmosphere, where it is plugging the leaky greenhouse blanket. On that basis alone, we expect that the temperature should have risen slightly over the past century. To calculate just how big a rise it should have caused, we have to consider some other effects. The atmosphere is sufficiently complicated that the computation is best done with a computer—a big computer.

Calculating Greenhouse Warming

The computer programs used to estimate global warming are very similar to the computer programs used to predict weather. They are very good, but they are limited in their ability to get the details right. The real complication comes from the complexity of the Earth and the intricacy of the flow of heat, air, and water. There are mountains and valleys, oceans and glaciers, snow and foliage. Energy is transferred not only by conduction and radiation, but also by transport—ocean currents and trade winds. Those can be modeled; more difficult is transport on the small scale: thunderstorms and hurricanes and dust storms. Worst of all is cloud cover. Clouds are highly variable, and they can cool or warm, depending on their thickness and altitude and the time of day. Heat is transferred not only vertically but horizontally, in ways we don't fully understand.

Everything is made more complicated by the response of the Earth to warming. A little carbon dioxide added to the atmosphere plugs the infrared leak and should certainly warm the Earth, provided nothing else happens. But other things do happen. Heating the oceans causes more water vapor to evaporate. Water vapor is also a greenhouse gas, so the temperature goes up even more. That's an example of positive feedback; you get more warming than you might have expected. Estimates vary, but calculations indicate that current water vapor feedback should approximately double the warming effect of carbon dioxide. On the other hand, more water vapor might increase cloud cover, which reflects sunlight and reduces the heating. That's negative feedback.

Why do I say increased water vapor *might* increase cloud cover? Amazingly, our poor understanding of cloud formation is responsible for the largest uncertainty in climate calculations. Clouds are complicated. They are patchy, they affect each other, their reflectance depends on their altitude and thickness, and they move. Sometimes they even lead to rain. All this is far too complicated for physicists to be able to calculate, even using the biggest and the best computers, so we resort to approximations and empirical relations from past experience. As a result, we wind up with huge uncertainties. That's why we can't be 100% certain that carbon dioxide increases the temperature. It is largely the uncertainty in the behavior of clouds that led the IPCC to conclude that there is a 10% chance that humans are not responsible for global warming. In this scenario, cloud cover is canceling the carbon dioxide effect, and the warming is due to an unknown natural effect, perhaps a continuation of our exit from the Little Ice Age. On the other hand, for most people, a 90% certainty that humans are responsible is high enough to demand action.

Another Danger: Acid Oceans

The increase in atmospheric carbon dioxide leads to another potential problem—one that worries some people more than global warming. About half of the carbon dioxide released into the atmosphere dissolves into the surface water of the oceans, and that makes the oceans slightly more acidic. We measure the strength of acids in units called *pH*, with lower pH meaning more acidic. The best estimate is that the pH of the oceans has already decreased by about 0.1 as a result of fossil fuel burning. If the carbon dioxide in the atmosphere doubles (and this is expected to happen by the middle of this century) the pH of the ocean surface waters will drop by about 0.23. By 2100, the total drop will be between 0.3 and 0.5, assuming that we burn fossil fuels at the expected (not treaty-regulated) pace. These numbers are far more certain than the predicted values for the temperature change.

Is such a pH increase bad? In fact, there is quite a bit of variability in the pH of the oceans right now—about plus or minus 0.1 pH for different locations. The expected increase in the acidity of ocean water is not as severe as acid rain, which has a pH lowered by 2 full units. In fact, right now the oceans are actually a bit alkaline (the opposite of acid), so the net effect will be to make the water slightly less alkaline and more neutral. But whether we call it acidification or neutralization, the specific concern is from the fact that dissolved carbon dioxide interferes with the formation of external skeletons and shells in many organisms, from plankton and algae to corals. A pH change of 0.2 or greater is likely to trigger noticeable changes in ocean life, and most people think such changes are unlikely to be good.

In a broader sense, the worry is that we are indeed now significantly changing the chemistry of the oceans. The pH of a liquid is very important in determining the rate of chemical reactions. We are experimenting with the oceans in a way that cannot be undone in any conceivable way in the foreseeable future.

The Ozone Hole

I include a few paragraphs here on the ozone hole problem because it is often confused with the greenhouse effect. You need to know the difference. Both the ozone and greenhouse problems have to do with pollution in the atmosphere and with the absorption of radiation invisible to human eyes. Other than that, the two problems are quite different. In fact, in many ways the ozone story is happier.

Sunlight consists of visible light, infrared heat radiation (IR), and ultraviolet light, or UV. Unlike IR, ultraviolet light plays no important role in the greenhouse effect, but it is the key player in the ozone problem. UV is also called *black light* and is used for Halloween displays because it is invisible to humans but can make some chemicals glow brightly. It is also the component of sunlight that does the most damage to your skin, causing sunburn and possibly cancer. UV light is so potent at killing bacteria that a black light is frequently used as a germicidal lamp for sterilization.

UV is dangerous because the individual photons carry much greater energy than do those of visible or IR light. In your skin, these photons can break apart DNA and cause mutations. In the atmosphere, they break up O_2 molecules into two individual oxygen atoms. These atoms attach themselves to nonbroken O_2 molecules to make O_3, also known as *ozone*. Ozone is a very strong absorber of UV radiation from the sun. This is another example of positive feedback: the air absorbs a little UV, and that creates a chemical (ozone) that absorbs even more. Most of the ozone is created between the altitudes of 40,000 and 60,000 feet, a region known as the *ozone layer*. The net result is good for us. The UV is absorbed in the upper atmosphere, and we are spared most of these deadly rays.

Without sunlight to create it, there is no ozone. That means that the ozone layer is absent over the South Pole during its long sunless winter. When the sun finally rises (once every year), the ozone layer forms. For decades scientists have been studying this ozone cycle using UV sensors in Antarctica. In the 1970s, they noticed that the amount of ozone formed was decreasing every year. This decrease became known as the *ozone hole*. Figure 20.4 shows a NASA plot of the growing ozone hole.

Was this ozone decrease natural or caused by man? Would the hole spread to the entire globe or be restricted to Antarctica? Nobody knew, although some people thought the hole might be due to a pollutant introduced into the atmosphere by humans. In fact, that is what it turned out to be. A chemical called *Freon* was in wide-spread use at the time—in refrigerators and air conditioners, and as a cleaning agent. Freon and its relatives contained the elements chlorine, fluorine, and carbon.[11] For that reason they are called *chlorofluorocarbons*, or CFCs. CFCs are highly stable; they don't decompose readily, so when they are leaked into the atmosphere from defunct refrigerators and air conditioners, they stay there a long time. CFCs are carried by winds and storms and eventually reach the ozone layer, where they are hit by ultraviolet light. The energetic UV photons break the CFCs into their constituents of chlorine, fluorine, and carbon atoms. It turns out that chlorine and fluorine are very effective at converting ozone back to ordinary oxygen, O_2. They are catalysts: they trigger the change but remain unchanged themselves, so they can keep acting over and over. Discarded refrigerators were, in effect, destroying the ozone layer.

The biggest effect happened to be over Antarctica. Nobody knew why, until atmospheric scientists realized that certain crystals of nitric acid formed there in the early spring, and on the surface of those crystals the chlorine and fluorine were far more effective at destroying the ozone.

Nobody knew for sure whether the destruction of ozone would continue until it reached more populated areas, but the world was sufficiently worried that it outlawed the use of CFCs in a treaty

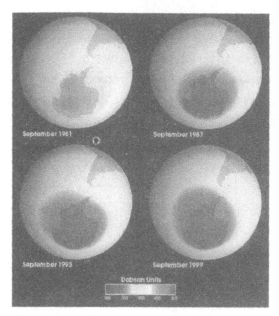

Figure 20.4. Ozone hole, 1981 to 1999.
Antarctica is prominent, and the southern tip
of South America is seen in the upper right.
Darker gray indicates ozone depletion.

called the *Montreal Protocol*. This agreement has been an outstanding international success. CFC production has dropped dramatically, and as a result, we expect the problem not to grow. The existing CFCs will remain in the atmosphere for a long time, but the situation has stabilized. However, the size of the remaining hole has been of continuing concern to citizens in Australia, since distortions in the shape of the UV window sometimes extend to the southern parts of their continent.

CFCs had also been used as a propellant for aerosol cans, for everything from shaving cream to insect repellent. It has been replaced for that purpose with other gases, including nitrous oxide. Some people still boycott aerosol products because they don't realize that the new ones are no longer dangerous to the ozone layer.

Because atmospheric chemistry is so complex, we don't know for sure whether the ozone hole would ever have extended beyond the

Antarctic region. Some people say that the real lesson from the ozone experience is that we can affect the atmosphere with human pollution, and that the effects are sometimes larger than we calculate. The success of the Montreal Protocol shows that international treaties can, in principle, be effective in stopping global pollution.

21

A Very Likely Cause

Global warming is real. Since 1957 the average temperature of the globe has risen about 1°F, and it is very likely that humans are responsible for at least part of that increase.

This claim seems surprisingly weak, given the stronger statements we often hear. Many scientists and politicians say, in effect, that humans are responsible for all of the warming of the last century. Is that true? In climate, it is hard to assign any causes with complete certainty. Let's look carefully at the IPCC 2007 conclusion. It says,

> The observed widespread warming of the atmosphere and ocean, together with ice mass loss, support the conclusion that it is extremely unlikely that global climate change of the past fifty years can be explained without external forcing, and very likely that it is not due to known natural causes alone.[12]

Some people criticize this statement as being too mild. They say that political pressures caused the human attribution to be weaker than it otherwise would have been. On the other hand, some scientists think the statement is too strong—that political pressure was responsible for exaggerating the human connection. I suggest that you, as a future president, take the statement at face value. It was approved by a very large committee of distinguished scientists. The words were carefully crafted. The statement is worth studying, because to the extent that there is a scientific consensus on global warming, this is it. So let's examine the words even more closely.

First let's examine the phrases *extremely unlikely* and *very likely*. The IPCC believes that the general public is uncomfortable with probabilities, so it uses those generic terms instead. But the report is written by scientists, and they were very careful (elsewhere in the report) to define precisely what they meant by their terms. Here is a glossary:

- *Extremely likely*: 95% chance, odds 19:1
- *Very likely*: 90% chance, odds 9:1
- *Likely*: 66% chance, odds 2:1
- *Unlikely*: 34% chance, odds 1:2
- *Very unlikely*: 10% chance, odds 1:9
- *Extremely unlikely*: 5% chance, odds 1:19

Using these definitions, I can rephrase the IPCC statement into more quantitative English:

> The observed warming from 1957 to now is extremely unlikely (5% chance) to result from ordinary climate variations. Something must have forced it to change (such as natural solar variability or human CO_2). It is very likely (90% chance) that humans are responsible for at least some of the warming.

Of course, this statement implies that there is a 10% chance that *none* of the observed warming is human caused. In its older 2001 report, the IPCC used the word *likely* (without the *very*) for human attribution, meaning 66% chance (that is, odds of 2:1)—so the evidence that humans are responsible for some of the warming became stronger in the subsequent 6 years.

In *An Inconvenient Truth*, Al Gore characterizes the claim that humans are responsible for global warming with these words: "The evidence is overwhelming and undeniable." He seems to suggest at several points in the film that humans are responsible for *all* (rather than *some*) of the warming of the past *100* (rather than *50*) years. He may be right—but it is worth pointing out that he is in disagreement with the IPCC consensus. The IPCC was very careful to define precisely what it means by its words. Al Gore takes a much more extreme position.

Here is a little game I've played that illustrates how political exaggerations misrepresent what the IPCC says. I tell a colleague I think there's a 10% chance that *none* of the global warming is caused by humans. Then I ask, "Based on that, do you consider me to be a *skeptic* on global warming?" The answer always comes back *yes*. But by that criterion, the entire IPCC is a global-warming skeptic.

Why are the IPCC conclusions so weak? Isn't the evidence for global warming really overwhelming, as Al Gore says? Yes, the evidence is very strong that global warming has happened. The IPCC gives it a 95% probability. But don't confuse global warming with *human-caused* global warming. These two ideas are related but not identical.

With the enormity of the data available, why can't the IPCC reach a stronger conclusion? The uncertainty comes down to basically one phenomenon: cloud cover. Virtually all the uncertainty in the climate model predictions comes from the model's poor ability to predict how clouds change when the temperature rises. The IPCC estimates that clouds exert a strong negative feedback, but not enough to completely cancel the warming. Moreover, natural cloud variations are not well understood. Clouds are affected not only by the presence of water vapor and by temperature, but also by variations in the sun and by the flow of electricity in the atmosphere. The computer models therefore base their predictions on empirical relations, rather than on relations derived from physics, making for even more uncertainty.

Doing Something About It: Alaska

Should we act to reduce carbon dioxide emissions, even though there is a 10% chance that they have no significant effect on global warming? Most people would say yes. If you were told that there was *only* a 10% chance that a terrorist was about to enter the room and blow it up, would you hesitate to leave because of the 90% chance that it wouldn't happen? Still, whatever action we take, it is wise to take it in a measured way. We want to do something that has a realistic probability of helping and is not just a symbolic gesture. The need to respond intelligently is illustrated by the warming of Alaska. To residents of this state, the prospect of even a small rise in average temperatures is a looming catastrophe. That's because Alaska is melting. Literally.

Much of Alaska is built on frozen ground called *permafrost*, a soil condition that results when the yearly temperature averages below freezing. But across most of the state, that criterion is just barely met, by a few degrees Fahrenheit. A small bit of warming can make a big difference.

When I drove Alaska's Highway 4 in the summer of 2003, the landscape looked flat but the ride felt like I was on rolling hills. The road undulated up and down, thanks to spotty drainage from partially melted permafrost; costly road repairs must be done every summer. Along the sides were "drunken trees" (a local term), leaning on each other's shoulders like thin, inebriated giants, their shallow roots loosened by soft soil. There were also drunken homes, leaning and sinking into the ground, and sunken meadows, 10 feet lower than the surrounding forest. Sunken meadows result when trees are cleared and a little bit of extra warmth reaches the ground in the form of direct sunlight.

The ecology itself seems to have a meltdown at about 32°F. Warm weather in Alaska encouraged an infestation of bark beetles that killed 4 million acres of spruce forest. This has been called the greatest epidemic of insect-caused tree mortality ever recorded in North America.

Alaska is frequently cited as the early-warning evidence that disastrous global warming is on its way. What can we do? Reduce CO_2 emissions to the 1990 levels? Our decision about what to do must depend on physics, and it's not as simple as some think.

In 2007, a startling paper was published showing that dust and soot from coal plants in China could be affecting the local climate of the Pacific Ocean enough to change the circulation patterns of wind and water.[13] If that is true, then it is conceivable that the melting of Alaska is due not to global warming, but rather to warm currents reaching the north from more temperate zones. And if that is true, then we could take immediate action to halt and perhaps reverse the warming of Alaska—by installing soot control at the Chinese power plants. If it is not true, and carbon dioxide is the real culprit, then the best thing to do in Alaska is to cope with the melting permafrost because nobody knows a practical way to remove carbon dioxide from the atmosphere, while taking measures to keep the CO_2 from increasing further. But even if we stop adding CO_2, the present warm temperatures will persist.

In other words, if the warming of Alaska is due to CO_2, then Alaskans must learn to live with it. If the warming is due to soot from China, then we can begin immediate negotiations with the Chinese to have them reduce their emissions. Our action depends critically on our understanding of the cause.

To make wise policy decisions, you must not only understand the science but also be familiar with the evidence. You must know not only the facts of global warming, but also which claims are exaggerated or distorted. It is important not to be misled by the pervasive misinformation that pollutes this field. The next chapter is devoted to straightening out the things that many people think are true but really "ain't so."

22

EVIDENCE

GRAPHS don't excite the public imagination. I have never seen the IPCC temperature plot (Figure 19.1) on the evening news, even though it is the most compelling evidence for global warming.[14] On the other hand, movies that show images of flooded New Orleans with distressed residents who have lost loved ones send a powerful and memorable message. As a result, most of the public discussion of global warming consists of dramatic anecdotes—everything from drought in Africa to polar bears who can't reach the receding ice floes of the Arctic.

The IPCC itself has been hesitant to make much use of these anecdotes, and with good reason. The linkage between these dramatic stories and global warming is speculative and perhaps wrong. Although the IPCC does use indicators other than temperature records (such as glacial retreat and rise of sea level), it most frequently uses the term *likely* to describe the possible linkage of these to human-caused warming, with that defined as 66% confidence. That is such a low confidence that it would not pass muster in scientific publications. In physics, the common standard is that you need to demonstrate 95% confidence in your result before a journal will allow you to publish it as a statistically significant observation.

What do you do if you are concerned that a disaster is imminent but the evidence is not strong enough to meet the usual minimum standards of science? Many choose to forget their normal criteria and show the public the heart-rending images of people suffering. Is this wrong to do? Perhaps it is valid to use the disaster of New Orleans as your example if you think of it as a metaphor for the horrors you sincerely believe are ahead, even though the link of that particular tragedy to global warming is tenuous.

The problem is that much of the public thinks such linkages are solid scientific results. In fact, much of what the public "knows" about global warming is based on distortion, exaggeration, or cherry picking. An example of *distortion* is the melting of the Antarctic ice—something that actually contradicts the global warming models but is presented as if it verifies them. *Exaggeration* includes the attribution of Hurricane Katrina to global warming, even though there is no scientific evidence that they are related. *Cherry picking* is the process of selecting data that verify the global-warming hypothesis, but ignoring data that contradict it. I'll give several examples in this chapter. Cherry picking is used in the US legal system. A prosecutor may present only the evidence that makes the accused appear guilty, and the courts depend on the defense counsel to put forth the contrary evidence. But even so, there are controls. Although not required to present contrary evidence in court, a prosecuting attorney is required to share all such information he possesses with the judge and the defense.

Physicists, by tradition, have a more stringent standard than the courts: if you are caught exaggerating, distorting, or cherry picking, your scientific reputation is damaged if not destroyed. I learned as a graduate student that in a presentation or a paper I am expected to present all the evidence—not just the truth but the whole truth, including any facts or analysis that might disagree with my conclusions. The public discussion of global warming rarely meets this standard of full disclosure. I think the reason is simple: when scientists present their case to the public, caution is interpreted not as good science, but as weakness. Among physicists, the opposite is true.

There is an irony here. The very reason that physics is so successful is that it holds a very high standard for determining truth. But in
global warming, because the issue is so important, the standard has
been lowered.

Yet I suspect the politicians are right. Muddy the waters, and the
public will procrastinate. Exaggerate, and the public is more likely
to accept your leadership. You are also far more likely to get news
coverage.

Exaggeration is a risky approach. When members of the public discover that they have been duped, they might react with backlash.
This happened in the mid 1980s in my home state of California, when
we suffered 6 years in a row of below-average rain and snowfall. At
the time, it was called the Great California Drought. Desperately worried about the future water needs for the expanding population of
the state, scientists and politicians exaggerated the dangers of the
drought, calling it *permanent*. Some scientists even suggested that the
water shortage was part of a pattern of *global warming*—a term that
had not yet worked its way into the wider public vocabulary. New
pipelines were built, water conservation was enforced, stricter standards for water use were introduced, and one coastal city, Santa Barbara, constructed an expensive ocean-water desalination plant. It was
a good investment, city officials thought, sure to be used forever.

Then, after 6 years, the rains came. They continued above normal for 4 of the next 5 years. Water from the desalination plant costs
about twice as much as water from reservoirs, so the system was
turned off. It still sits there, in standby mode, waiting to be used for
future droughts.

In fact, California *is* short of water, and it will be short of water for
the foreseeable future. The drought provided a good opportunity to
bring that problem to the public attention. But by exaggerating the
case, scientists and politicians lost credibility. Not all had issued the
dire warnings, but the ones who had were the ones who made it into
the newspaper headlines. The public felt that the scientists and
politicians had cried wolf. Many of them had.

Can a similar thing happen with global warming? I fear it might.

Even if global warming is caused by human burning of fossil fuels, we could easily have another 5 years of constant temperature, perhaps even cooling, just from natural variability. If, 5 years from now, we have had 15 years of temperatures cooler than those in 1998, what will be the public response? As a future president, you must consider this issue very carefully. Will the climate crisis community be discredited? Will the public pay attention to you when you argue that fossil fuels are still dangerous and the recent cool weather was only a fluctuation?

Let's examine some of the anecdotal evidence that has been used to excite the public about global warming. But keep in mind, showing that evidence in support of global warming is false does not imply that global warming doesn't exist, or that it is not caused (at least in part) by humans.

Distortion

In 2006, scientists reported the results of a remarkable study. Using a pair of satellites called GRACE, they were able to measure changes in Antarctic ice from 2002 to 2005. They did it by detecting the small but clear changes in the gravity of the ice, as I described in Part IV. It was a remarkable achievement and is undoubtedly the most accurate estimate of the ice changes ever made.

The results of this study were scary: Antarctica was losing 36 cubic miles of ice every year. Note that that's cubic *miles*, not cubic meters. Even a huge city like Los Angeles uses only 1 cubic mile of water each year. Isabella Velicogna, one of the authors of the report, stated, "This is the first study to indicate the total mass balance of the Antarctic ice sheet is in significant decline."

This melting of Antarctica was widely reported as strong new evidence in favor of global warming and indicating the dangers that we will face if it continues. If all of Antarctica melts, the sea level will rise over 250 feet, flooding most of the world coastline, including New York City.

In fact, however, the melting-Antarctica result didn't verify the global-warming predictions but actually contradicted them. As the authors of the report made clear, the previous IPCC report in 2001 had predicted that global warming would *increase* the Antarctic ice mass. It is easy to see why this paradoxical prediction makes sense: Global warming causes increased evaporation of the ocean waters. When this extra water vapor reaches Antarctica, it falls as snow—because even with the present 1°F of global warming, most of Antarctica remains well below freezing. More snow means more Antarctic ice, not less. Yet the opposite is observed.

Does the decrease in ice mean that the model is wrong—that global warming is not taking place? No, not at all. It simply shows the inadequacies of the model. Even with global warming, local weather (even for a whole continent) can cause behavior that deviates from the computer calculation. One result is certain: the melting of Antarctica provides no evidence whatsoever in favor of global-warming predictions. To use it as such evidence is not science, but propaganda.

Another example of distortion is found in the claim that the cost of hurricane damage has been rising exponentially because of global warming. The plot in Figure 22.1A illustrates that rise. This plot has been widely used, and a version of it appears in *An Inconvenient Truth*. It is highly misleading, however, because it shows costs without compensating for inflation. The old costs of hurricanes seem very low, but that impression is, in large part, because those old dollars are shown at only a few percent of their true value, measured by the income of those who lived at the time. Moreover, if we compensate both for inflation and for the increased number of people living near the coast (and the increased values of their homes), we get the plot in Figure 22.1B. This revised plot shows that the two big hurricanes of 2005—Katrina and Wilma—were no more costly than the 1928 Okeechobee hurricane. Hurricane costs vary a lot, but there is no obvious trend up—or down.

Many people are shocked when they learn that the widely cited plot in Figure 22.1A doesn't compensate for inflation. In any under-

A

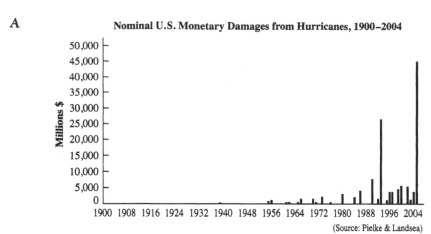

Nominal U.S. Monetary Damages from Hurricanes, 1900–2004

(Source: Pielke & Landsea)

B

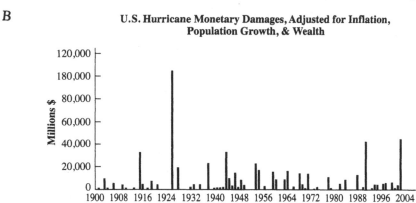

U.S. Hurricane Monetary Damages, Adjusted for Inflation, Population Growth, & Wealth

(Source: Pielke & Landsea)

Figure 22.1. Monetary damages from US hurricanes. The plot in (A) shows the dollar cost from 1900 to 2005. The plot in (B) shows the cost when it is compensated for inflation and the increase in building near the coast.[15]

graduate course, ignoring inflation would earn the student an immediate F. Yet in public discourse, such distortion is routine. It makes for more compelling news—not just for global warming, but for many other things, from oil and gasoline prices to box office receipts for new movies. As president, you must absolutely insist that every chart be plotted in constant dollars. Figure 7.2 illustrates that such adjustment for inflation also affects our incorrect perception that gas and oil prices are historically much higher than ever before.

As a final example of distortion, consider the dramatic image of polar bears that died because they were unable to find an ice floe—again, an image that is dramatically portrayed in *An Inconvenient Truth*. The issue was brought to a British judge by a school governor who objected to the widespread distribution of the movie to school-children. After reviewing the published results on the subject, he concluded, "The only scientific study that either side brought before me can find is one which indicates that four polar bears have recently been found drowned because of a storm."

When I visited Alaska to see the melting of the permafrost first-hand, I found that scientists there were well aware of the public disinterest in numbers and plots. They sarcastically referred to bears and other large, cute animals that drew public interest as "charismatic megafauna."

Exaggeration

Some climate models suggest that increased warming could cause increases in storm activity such as hurricanes, brought about by the increased energy available in warmed tropical oceans. However, the models are not very good at simulating storms, and some scientists argue that storm activity should not increase but actually decrease. Their claim is also based on sound reasoning. Virtually all global-warming models predict that the polar regions will be warmed even more than the tropics. Storms are generated and intensified when big temperature differences occur, so reduction of temperature gradients should prevent storms from growing. Which analysis is correct? We don't know.

Given this uncertainty, what should we do? One possibility is to recognize that increased storm activity is a possible consequence of global warming, and such a result could be very bad. So caution dictates that we act as if that prediction were true. But is that really the cautious approach? What if global warming reduces storms?

When theory fails, look at the record. The results have been presented on TV, in *An Inconvenient Truth,* and in numerous articles and newspaper reports. Here is what most people know from these sources: the evidence is overwhelming that the number of truly intense hurricanes, those in categories 4 and 5 (winds over 160 miles per hour), has increased dramatically in the past 30 years. In 2005 alone, we set a record with five such storms. Katrina was only the worst.

If that is what you know, it is important that you learn that it *ain't so.*

The most frequently cited scientific investigation on the subject was published in *Science* magazine in 2005 by P. J. Webster and his colleagues.[16] Their report was thorough and careful. They did show that the number of observed category 4 and 5 storms more than doubled between 1970 and 1990. That is the foundation of most of the fear, but that isn't the whole truth. They also showed that the number of such hurricanes has not increased since then. (It has actually decreased slightly, but not by an amount that is statistically significant.) The authors were concerned that part or all of their observed increase could be attributed simply to our improving ability to detect and measure storms that are out at sea, and in their paper they warn against interpreting the result as a true increase.[17] Their warning was not heeded by many people.

Subsequently, in 2007, an analysis by C. W. Landsea[18] analyzed a particularly strong bias in the data, based on the fact that our means for detecting hurricanes and their strength had changed over the years. Formerly, we discovered hurricanes only when they hit a shipping lane or the coast; now we detect them far out at sea using satellites and automatic buoys. Landsea showed that the increase in hurricane numbers since the 1930s is due largely to our improved early-detection equipment.

There is also a bias in the way storms are reported. Hurricane Katrina was labeled category 5 only because its maximum wind speeds were tracked continuously during its evolution. In fact, it was only a category 3 storm when it destroyed New Orleans.

Fortunately, scientists have developed methods to eliminate such biases. One good way is to plot only those hurricanes that actually strike the US coast, since these are discovered with 100% probability. This approach—finding an unbiased subset—is widely used in physics as a straightforward method for avoiding biased conclusions from biased data. The number of hurricanes that actually made landfall is plotted in Figure 22.2.

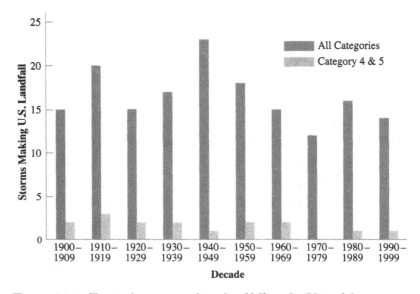

Figure 22.2. Tropical storms making landfall in the United States. The tall bars give the number of all storms in each decade; the shorter bars represent the number of storms in categories 4 and 5.[19]

The figure shows no obvious trend up or down in the number of hurricanes hitting the United States, or in the number of intense (category 4 or 5) hurricanes. Because the current decade is not complete, the plot shows no data for the years 2000 to 2009. Isn't that unfair? After all, many people remember 2005 as the year of severe hurricanes, when five category 4 and 5 storms hit the US coast: Dennis, Emily, Katrina, Rita, and Wilma. That will make the newest bar (the small one) set a super record high, won't it? No. The high wind speeds from those hurricanes never hit the United States; they all occurred out at sea. All five of those storms were category 3 or less when they actually struck land. In fact, from 2000 to 2007, only one storm hit US land while it was still category 4 or 5: Hurricane Charlie in 2004. So the observations through the end of 2007 suggest that the plot is unlikely to change very much when the newest decade is added.

Experts attribute the strong storms of 2005 to the fact that the Caribbean was unusually warm that summer, by several degrees, because of an El Niño event. Couldn't that increased water temperature have been due to global warming? We can't rule that out, but there is no evidence supporting that contention. Moreover, New Orleans was not struck by a category 4 or 5 storm, and any category 3 storm that hit the city in the past 50 years could equally well have destroyed that city.

Recall that global warming has its greatest effect in the polar regions, not in the tropics. In the last 50 years, the IPCC says, the temperature rise has been about 1°F. Changes in Caribbean water temperature are strongly driven by the El Niño cycle, and that cycle is widely believed to be responsible for the storms of 2005. The linkage between global warming and Katrina is a hypothesis, not a scientific result.

After Katrina, many politicians and scientists warned that this was the beginning of a period of terrible damage from more hurricanes. Then, the following year (2006), El Niño shifted, and not a single hurricane struck the US coast. That news, or the absence of it, hardly made the papers.

Cherry Picking

If you track aspects of climate, many of them will change—some for worse and some for better. If you pick only the bad changes, you can compile a scary collection.

Consider tornadoes. In *An Inconvenient Truth*, Al Gore also states that the number of tornadoes has been increasing and attributes that worrisome fact to global warming. In fact, however, the study published by the US government suggests otherwise. Every year the US National Oceanic and Atmospheric Administration (NOAA) publishes a report in which it plots the record of tornadoes and other violent events.[20] The recent NOAA report "The Climate of 2006" contains a plot of the number of strong to violent tornadoes in the United States from 1950, shown in Figure 22.3.

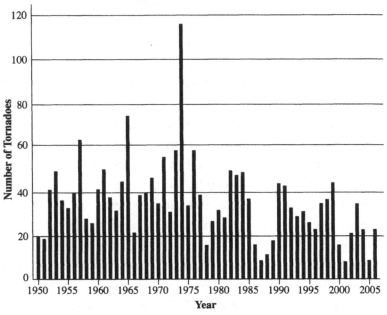

Figure 22.3. *Number of strong to violent (F3–F5) tornadoes in the United States (March to August), 1950–2006.*

The figure suggests (and a detailed analysis confirms the impression) that the number of strong to violent tornadoes has been slowly decreasing over the past 55 years. So how could Gore claim that the number is increasing? The answer is that for tornadoes (unlike hurricanes), he did not cite the number of strong to violent storms, but only the total number detected. In fact, the number of detected tornadoes is going up, probably because of improved radar to detect small and distant tornadoes. The number of big ones—the ones that do damage—is actually decreasing. That sounds good, so why mention only the total increase, which sounds bad? Possibly because by cherry picking the data, someone who is deeply concerned about the dangers of global warming can grab the attention of the public.

Wildfires provide another example of cherry picking. Al Gore states that the number of wildfires has been increasing, and he attributes that increase to global warming. It turns out that the annual NOAA report, cited earlier, also includes plots for wildfires in the United States, as reproduced in Figure 22.4. Part A of the fig-ure shows that the acres of burned land are indeed increasing, but

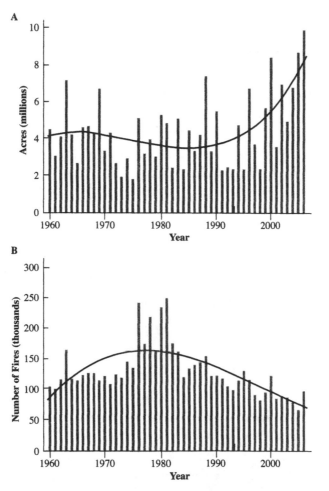

Figure 22.4. NOAA data on US wildfires, 1960–2006. (A) Number of acres burned in wildfires. (B) Number of wildfires.

part B shows that the number of fires is actually decreasing, contrary to Gore's claim. However, these different claims do not really contradict each other. It turns out that Gore was referring not to acreage or number, but to "major disasters," not just in the United States but throughout North America. Major disasters are measured by human consequences, and these increase as people move closer to the wilderness. The NOAA reports of a decrease in the number of fires and an increase in acreage could both be related to policy to let fires in the wilderness burn themselves out.

Do the plots show good news or bad news? Both. If you want to scare the public, find a plot that sounds awful; if you want to reassure, choose a different one. That's cherry picking.

For another look at cherry picking, let's return to the example of the melting of Alaska. In Chapter 21, I described the awful effects of the melting permafrost, and the possibility that the cause might not be global warming but a local effect. We don't know. Now let's look at the temperature record published by the Alaska Climate Research Center for Fairbanks, Alaska (Figure 22.5).

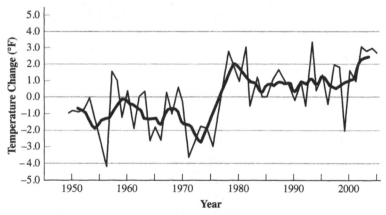

Figure 22.5. Temperature of Fairbanks, Alaska, 1949–2005, reported by the Alaska Climate Research Center.

The figure shows that the warming of Alaska is real and documented. Anecdotes of houses sinking in melting permafrost aren't the only indicator; this plot shows hard evidence. Yet there is something peculiar about the plot. The warming did not take place gradually but happened rather suddenly, between 1970 and 1980. Moreover, the temperature has been relatively constant since then. Compare this plot to the global-warming graph (Figure 19.1). The patterns of warming in Alaska and on the Earth as a whole seem quite different. Most of global warming has taken place in the past 28 years, from 1980 until now. Yet the temperature of Alaska has been very stable in that same period.

Does the strange pattern show that the melting of Alaska is not due to global warming? No, not at all. The temperature trends of Alaska could well consist of a rise due to global warming, with a downward fluctuation in the last decade caused by something else. (Maybe Chinese soot is cooling Alaska!) However, advocates of action are unlikely to show you this temperature plot, because it raises awkward questions about the cause of the melt. That's another kind of cherry picking. Show only the data that wows the audience, and avoid anything that seems to contradict the simple picture. But scientists who are trying to figure out real causes have to see all the evidence.

News Bias

We are bombarded with news purporting to show that global warming is real, imminent, and disastrous. Be suspicious; remember that global warming amounts to 1°F in the past 50 years; anything larger than that is probably a local effect being misrepresented as global. As an example, consider an op-ed piece that appeared in the *New York Times* on June 28, 2007, written by Nicholas Kristof and titled "Our Gas Guzzlers, Their Lives." The article tells of hardship in Burundi brought about (he claims) by human-caused global warming. Kristof says that changing weather patterns have already caused crop failures. He mentions the 50-foot retreat of the shores of Lake Tanganyika. Lake Victoria was dropping by half an inch per day.

Could all this be due to the 1°F temperature rise? Actually, the temperature changes near the equator are typically less than near the poles—and 1°F is the average. So the real question is, Could all of this be due to a ½°F temperature rise?

The answer is clearly no. Weather and local climate vary greatly. You can always find a location that is suffering (for example, Alaska or Burundi) and attribute the effects to global warming, but only if you abandon scientific care. If you find an area that is extra cold, you can blame the low temperature on the wider variability that some people argue should accompany global warming. If you find that an area is extra warm, you can blame that, too, on global warming. You can blame everything on global warming. But if you engage in that kind of cherry picking, you are really fooling yourself. If global warming were so obvious, the IPCC would not have stated that there was even a 10% chance that humans are having no effect.

Does all this mean that we should ignore the tribulations of the people of Burundi? Of course not. But it suggests that the way to help them is by sending aid, not by trying to reduce CO_2 emissions.

Cherry picking in order to inspire people is a political ploy that must be used with caution. Be sure never to fool yourself into believing your own exaggerations, and be prepared for possible backlash when people discover that they have been misled.

The Hockey Stick

The hockey stick story that I'm about to tell illustrates the danger when you see something that seems to support your prior belief. There is a strong psychological pressure to drop your normal skepticism, to accept the new result, and rejoice that your intuition is now validated. The story of the hockey stick is a tale that every future president needs to know.

The *hockey stick* is the nickname for a remarkable graph that too quickly became a "poster child" for the environmental movement. Plots published by Michael Mann and colleagues in 1998 and 1999 purported to show that the climate of the Northern Hemisphere had been remarkably constant (and cool) from AD 1000 to 1900, until global warming began to heat things up. The overall shape of the curve resembled a hockey stick lying on its back—a straight part representing prior constant climate, with a sudden bend upward in the twentieth century due to global warming.

The hockey stick was turned from a scientific plot into the most widely reproduced image of the global-warming discussion. The version shown in Figure 22.6 is based on the 2001 report of the IPCC. The hockey stick figure appears five times in just the summary volume alone. It is one of the prominent plots that Vice President Gore shows in *An Inconvenient Truth*.

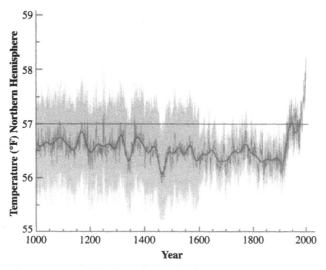

Figure 22.6. The hockey stick plot, an attempt to bring together all the best records to give a true global average of temperature over the past 1000 years. The name derives from the resemblance of the shape to a hockey stick. The gray regions show the range of uncertainty.

What made the hockey stick plot so interesting was that it purported to represent the entire Earth with less bias than any prior plot. What made it so important was that it clearly showed that the modern fossil fuel era, the twentieth century, was unprecedented. The shape of this plot matches the shape of the carbon dioxide plot

(Figure 20.3). The similarity seemed to make it patently obvious that humans were the cause of global warming. The year 1998 (when the hockey stick was first published) was clearly warmer than any other of the last 1000 years, even if the uncertainties of past climate (the gray area in the plot) are included.

The medieval warm period, whose end (according to Barbara Tuchman) led to a physical chill in the fourteenth century and subsequent miseries, appears nowhere in this plot. Mann and his coauthors concluded that its absence showed that this supposedly global warm period had been merely European in extent. The implication was that Eurocentrism had biased previous scientists into the simple but understandable error of assuming that their tiny continent represented the whole world. In *An Inconvenient Truth*, Al Gore has fun mocking the medieval warm period by pointing to that region where some scientists claim it exists, while the audience laughs because of its obvious absence in the hockey stick plot.

The hockey stick was so dramatic that the Canadian government distributed it to every household in Canada to gain support for ratifying the Kyoto accord limiting greenhouse gas emission. Virtually every politician in Washington DC was familiar with the plot. It became the most beloved figure of many anti-global-warming activists. The accompanying sound bite was that, thanks to humans, the climate on Earth was now the warmest that it had been in a thousand years.

Then came a shock. Canadians Stephen McIntyre and Ross McKitrick uncovered a fundamental mathematical flaw in the computer program that had been used to produce Mann's hockey stick result. In his original publications of the hockey stick, Mann said he had used a standard method known as *principal component analysis*, or PCA, to find the dominant features in a set of more than 70 different climate records. But it wasn't so. McIntyre and McKitrick obtained a key part of the computer program that Mann used, and they found serious problems. Not only did the program not do conventional PCA, but it handled data normalization in a way that can only be described as wrong. Mann's improper normalization procedure tended to emphasize any data that had the hockey stick shape, and to suppress all data that did not. In fact, instead of representing the worldwide record, the hockey stick was primarily a history of the climate of the western United States. Mann's hockey stick plot was not more universal than the previous work, not more free of bias, but just the opposite.

This error was very embarrassing for people who had publicized the hockey stick as the smoking gun of global warming. Congress requested that the National Research Council of the US National Academy of Sciences conduct a review. I was asked to be a referee on the review. After considerable work, the report concluded that the strongest statement that could be made was that the present years were the warmest in the last 400 years, not 1000 as Mann had said. Of course, that fact wasn't news. It was known even back in 1990, when the IPCC issued its first report. The council also concluded that Mann's analysis showing the absence of the medieval warm period was not correct (despite Al Gore's sarcasm). Its report stated that Mann had underestimated his errors (the gray regions in the hockey stick plot), and that with the correct and much larger error bars, no conclusion about the presence or absence of that period could be drawn from Mann's data. In the end, there was nothing new left in Mann's papers that the National Academy supported, other than that the idea of using principal component analysis was, in principle, a good one.

Does it matter that the hockey stick was wrong? No, not if you are a scientist. The IPCC consensus did not depend on the hockey stick. But it is embarrassing if you are a politician who has been telling constituents that the climate is the warmest it has been in a millennium and now you have to change your story. This result should not affect any of our thinking on global warming, but it is a cautionary tale to politicians in general and to future presidents in particular. It is not wise to jump on the bandwagon of a new report simply because it verifies what you previously believed. Scientists usually reach the truth, but sometimes it takes a while.

Carbon Dioxide and Ancient Climate

For the final example of this chapter, I reproduce in Figure 22.7 a plot from *An Inconvenient Truth*. The plot is not what it seems. Vice President Gore says nothing inaccurate about it in the film. Yet most people come away from the movie with an incorrect conclusion. The example illustrates how easy it is to misinterpret important results.

The plot shows ancient records of atmospheric carbon dioxide, recovered from cores going back 600,000 years. Right below that graph, Gore plots the record of ancient temperature, with huge dips representing the recurring ice ages. The two graphs are strongly correlated. In fact, they seem to track each other extremely well. Almost every nonexpert who has seen the movie comes away believing that these changes in carbon dioxide were responsible for the ancient ice ages. Yet Vice President Gore never actually says that. These are his words:

> Here is an important point. If my classmate from sixth grade were to see this—you remember, the guy who asked about South America and Africa [continental drift]—he would ask, "Did they ever fit together?"
>
> The answer from the scientists would be, "Yes, they do fit together."
>
> It's a complicated relationship, but the most important part of it is this: When there is more CO_2 in the atmosphere, the temperature increases because more heat from the sun is trapped inside.
>
> There is not a single part of this graph—no fact, date, or number—that is controversial in any way or is disputed by anybody.

Almost everyone who sees this movie regards the plot in Figure 22.7 as one of the most dramatic pieces of science they have ever seen. They interpret it as conclusive evidence that carbon dioxide drives

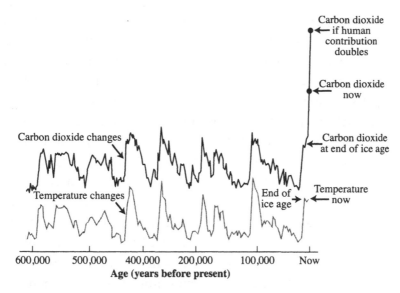

*Figure 22.7. The ancient records of temperature and carbon dioxide,
showing the correlation, adapted from* An Inconvenient Truth.

climate. Yet many climate scientists think the opposite: that the large
changes in carbon dioxide were a *result* of the warming, not the cause.
That doesn't mean that the carbon dioxide changes didn't contribute
a little bit to the warming, but we can estimate how much they con-
tributed by looking at the right-hand part of the plot. The first big dot
shows the level of carbon dioxide in the present atmosphere. According to our best estimates, from both physics calculations and the large
global climate models in the computers, that much of an increase of
carbon dioxide should result in a temperature rise of about 1°F. How
much did the carbon dioxide change during the ice ages? The answer
is revealed in the plot: by about the same amount as the current
increase. So we would have expected such changes to cause 1°F
changes in climate. In fact, the changes are more like 10°F or 15°F.
How could they be so large? The likely answer: the temperature
changes were not driven primarily by carbon dioxide changes, but by
changes in the orbit and axis of the Earth.

Additional evidence suggests, for the ancient data shown in Figure 22.7, that the climate is changing the carbon dioxide, rather than

the other way around. The evidence comes from a study published in *Science* magazine in 2003[21] arguing that the climate changes occurred *first*, and then were followed by the carbon dioxide changes. The CO_2 lag is 800 years, with an uncertainty of only plus or minus 200 years. This result is based on the observation of changes in atmospheric argon gas. It is not fully understood if these changes reflect temperature, but it is clear that dramatic changes were occurring on Earth 800 years prior to the carbon dioxide changes.

The CO_2 variations are not new; they have been part of standard undergraduate courses on the ice ages for decades. In those classes, students were taught that the CO_2 was responding to climate, not the other way around. The reason is that CO_2 gas is more soluble in cold seawater than in warm. When the climate warms, carbon dioxide dissolved in the oceans escapes into the atmosphere and increases the concentration in the air. When it gets colder, the carbon dioxide in the air tends to dissolve in the seawater.

Why is there an 800-year lag? Because it takes that long for the deep ocean water to circulate to the surface where it can release its carbon dioxide. So, for example, consider the beginning of a warm period. The warm surface water emits some of its dissolved carbon dioxide into the atmosphere, but the deep ocean water doesn't. It just isn't in contact with the air. After about 800 years, virtually all of the ocean water has reached the surface, warmed (at least temporarily), and released its carbon dioxide.

Some climate scientists now doubt this old explanation and argue that the plot shows that CO_2 is the real cause and climate is the effect. It is an interesting scientific debate, and one in which I have taken part. The fairest statement that can be made is that the issue is unsettled. That's probably why Vice President Gore was so careful in what he said. Read his words again. He states that the relationship is "complex." He doesn't take the time to explain what I just explained, possibly because not all of the general public has the same intense interest in this subject that you have (and you wouldn't have read this far if you didn't). Many people see the plot in Figure 22.7 and come away believing that carbon dioxide has been established as the main driver for past climate change. That is "very likely" a correct conclusion for small (1°F) changes attributed to the present CO_2 increase, but not clear for the dramatic temperature variations in the ancient data shown in the plot. When people learn of the uncertainty, however, they should not draw the wrong conclusion that carbon dioxide doesn't matter. Sometimes when people feel they have been duped, they overreact. You may have to reconvince them of the reality of global warming and its real dangers when you become president.

Science and Propaganda

A colleague of mine told me that in a poll asking the public to name a living scientist, the person whose name appeared most often was Al Gore. I don't know if this is true or apocryphal, but it is plausible. Gore took a class from the great scientist Roger Revelle, the original guru of global warming (and a longtime friend of mine, until he passed away in 1991). He talks all about his scientist friends and what they tell him. In *An Inconvenient Truth*, he does come across as a scientist, but I am sure that he would not claim that distinction himself. And because he is not a scientist, he can be much more effective as a public spokesman because he doesn't have to abide by scientific standards.

Al Gore won the Nobel Peace Prize because of his effectiveness in alerting the US public to the dangers of carbon dioxide and global warming. He was able to do this through a combination of artistry, powerful writing, and exaggeration, mixed with some degree of distortion and a large amount of cherry picking. His great movie *An Inconvenient Truth* is powerful propaganda, but as with all propaganda, there is a danger. When it is discovered that Gore has exaggerated the case, the public may reject the truly scientific case for fossil fuel–induced global warming. To use an old cliché, I fear that the public will throw out the baby with the dirty bathwater.

23

Nonsolutions

GLOBAL warming is real. It is very likely caused by humans. By the end of the twenty-first century, it will (if caused by humans) grow enough to be disruptive. What can we do about it? What can *you* do when you ascend to the presidency? With China now the leader in greenhouse gas emissions, the problem seems hopeless. Chinese leaders correctly state that their citizens have the same right to a high standard of living that we have, and they believe that achieving that standard requires energy. History bears them out. Here's the problem in a nutshell:

- The cheapest and most readily available source of energy is coal.
- Coal produces more carbon dioxide per kilowatt-hour of energy than virtually any other source.
- China has enough coal to last more than a century, even with the expected increases in economic growth.
- China is currently building over 1 gigawatt of new coal plants every *week*.

Do we demand that the Chinese stop? Do we have the right to do that? Do we have the power to do that? I believe the only reasonable

300

answers to those questions are no, no, and no. It gets worse. Provided that the economy of India continues to grow, it too will soon over-take the United States in emissions of greenhouse gases. Of course, we all hope that India's economy does grow. And we hope the same for China and Russia.

Figure 23.1 shows carbon dioxide emissions from various regions of the world during the last century. We in the United States enjoy a wonderful standard of living, and the rest of the world deserves that too. Look at the plot and imagine that we stop the US growth in carbon dioxide emissions. In fact, let's reduce them by 20%, from the current level of about 1.5 billion tons to 1.2 billion tons. How much will that help? If China and India continue to increase their production at the current rates, they will cancel our cutback in just 3 years. That means that a permanent reduction in US carbon diox-ide emissions, down 20% from our level in the year 2000, will delay global warming by only 3 years, nothing more. Delaying the problem doesn't solve it. Is the situation hopeless?

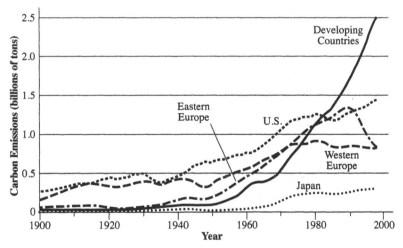

Figure 23.1. Carbon dioxide emissions in the twentieth century.

Actually, I am optimistic. Hidden beneath the hype are some solutions, but they are obscured by many nonsolutions. I'll use physics to show that a lot of the commonly believed approaches will not work. The problem for any future president will be to guide the public in a meaningful direction. Given the huge amount of things they know that ain't so, that won't be easy.

In this chapter I'll begin with the things that many people hope to use but that will probably not work.

Hydrogen Hype

Many people believe that hydrogen is the best fuel source for the future. Pound for pound, it has 2.6 times the energy of gasoline. The only waste produced is water—no pollutants, no carbon dioxide. In his 2006 State of the Union address, President George Bush expressed his hope that our children would someday be driving hydrogen automobiles. Governor Arnold Schwarzenegger of California immediately endorsed the plan and directed his state to start building the infrastructure needed to make it work. Problem solved? No.

After just a few months, Governor Schwarzenegger withdrew from the hydrogen hype. Perhaps even more amazing, in his 2007 State of the Union message, President Bush made no mention whatsoever of the hydrogen economy—a startling and important fact that was missed by virtually all commentators on the speech. Why did two of our most important leaders change their minds? What happened? I think the answer might be that they both learned some physics.

I'll repeat here some of the hydrogen physics that I discussed in the chapter on key energy surprises. Hydrogen does indeed have 2.6 times the energy of gasoline per pound—but because hydrogen is so light, a pound of it takes up a lot more space. The gas can be compressed, but only if it is put in strong and heavy containers, which negates the weight advantage. When cooled to −423°F, hydrogen becomes a liquid that can be stored, at least for a few hours, in a thermos-like container; but even as a liquid, hydrogen is still very light. A 10-gallon auto gas tank will hold 62 pounds of gasoline, but only 5.7 pounds of liquid hydrogen. Here are two facts that a president needs to know: as a liquid,

• Hydrogen has about *three times more energy per pound* than gasoline[22]
• Hydrogen has about *three times less energy per gallon* than gasoline

Let's look at some examples. Suppose you get 30 miles per gallon in your present car. If you filled the tank with liquid hydrogen, you would get only 10 miles per gallon. The car range would be three times shorter than you have now. If, instead of liquid hydrogen you used compressed hydrogen gas (with a heavy tank to contain it), you would get between 2.5 and 5 miles per gallon, depending on the thickness of the tank walls. Let's apply these numbers to ranges. If your current car has a 15-gallon gas tank, then for equal range you would have to carry 45 gallons of liquid hydrogen, or 90 to 180 gallons of compressed hydrogen gas. You could also keep your present-sized tank and try to live with a shorter range. Fifteen gallons of compressed hydrogen gas could take you only 12 to 45 miles between fill-ups.

Don't be surprised to read articles with titles like "Who Killed the Hydrogen Economy" that blame the oil companies. It wasn't them. It was the physics of hydrogen.

Despite these numbers, hydrogen has potential uses. It could be used in large vehicles, such as buses and trucks, that can afford the large tanks and frequent refill needs. Hydrogen might be the fuel of choice for an ultralight airplane, since weight in such vehicles is very important. Not only is the hydrogen light, but the fuel cells and electric motors to use it can also be light. Unfortunately, these are all rather specialized uses, and their implementation will have no major impact on the carbon dioxide problem.

There is nevertheless some hope. Hydrogen could be useful for automobiles, if we could improve the mileage significantly, perhaps by making the cars much lighter in weight. If autos could get 100 miles per gallon with gasoline and 33 with liquid hydrogen, they could have a 330-mile range with only a 10-gallon tank.

Another problem with the hydrogen economy is that there are no hydrogen mines or other sources of the fuel. Hydrogen is not harvested; it is manufactured. We can obtain hydrogen by electrolyzing water—that is, by breaking up H_2O into its constituents of hydrogen and oxygen. But that takes energy, yielding no more than is put in, and typically less. So think of hydrogen like this:

Hydrogen is not a source of energy.
It is only a means of transporting energy.

Electrolysis of water is not the cheapest way of making hydrogen. In the United States, hydrogen is usually produced by reacting methane (natural gas) with steam—a process called *steam reforming*. Coal can be used in place of methane. These processes also create carbon monoxide, which is burned to carbon dioxide, so they are not clean. Moreover, you might obtain more energy, and no more pollution, by simply burning the fossil fuel in the first place. That doesn't mean that making hydrogen isn't worth doing. By converting the fossil fuels at the hydrogen manufacturing plant, at least you have the option of sequestering the carbon dioxide—that is, burying it deep underground, which you cannot conveniently do with the CO_2 emitted by your auto. I'll discuss the carbon capture and sequestering option (CSS) in the chapter on new technologies.

Electric Automobiles

Electric automobiles sound like a good idea. The news is full of hyperbole about the great new Tesla Roadster and plug-in hybrids. Charge your car by plugging into the wall, and drive around with zero emissions. The Toyota Prius is half-electric, and that's been a big success, right? Moreover, wall-plug power is cheap. A kilowatt-hour costs only 10 cents. At $3 per gallon, the cost of energy from gasoline is about 36¢ per kilowatt-hour. That means that recharging a car by plugging it into the wall socket is equivalent to buying gasoline at under a dollar per gallon!

Alas, electric automobiles have very serious problems. The fundamental one is that batteries store very little energy, compared to gasoline. High-quality, expensive batteries—the kind used in cell phones and portable computers—store only 1% as much energy as gasoline, pound for pound. That's a big factor. It is offset somewhat by the fact that electric energy can be used more efficiently than gasoline, so the disadvantage for batteries is actually more like a factor of 30. For the same range, you can carry 62 pounds of gasoline, or 1860 pounds of batteries—almost a ton. Because batteries have more pounds per cubic foot than gasoline, they won't take up 30 times the space—but only 10 times as much. That's why cars like the Tesla Roadster are possible.

The cost savings are illusory. High-performance batteries are very expensive and need to be replaced after typically 700 charges. Here is a simple way to calculate the numbers. The computer battery for my laptop (on which I am writing this) stores 60 watt-hours of electric energy. It can be recharged about 700 times. That means it will deliver a total of 42,000 watt-hours, or 42 kilowatt-hours, before it has to be replaced for $130. Put those numbers together to get the battery replacement cost: $130/42 = approximately $3 per kilowatt-hour. That's 30 times more expensive than the 10¢ per kilowatt-hour to charge it. The real expense for fancy batteries is not the cost to recharge them, but the cost to replace them. The same factor will be true for the Tesla Roadster. Driving it will seem very cheap, until the time comes to get new batteries. They are by far the most expensive component of that car.

Can't batteries be made cheaper? Yes, but then they become unreliable and typically take fewer charges. That has probably already been your experience if you've purchased Chinese copies of expensive batteries for your computer, cell phone, or digital camera. The reason for the high cost is, in part, that batteries are intrinsically dangerous. Poor quality control leads to batteries that spontaneously burst into flames. Engineers are working hard to solve these problems.

Can't the technology be improved to allow more recharges? Yes, and engineers are working on that. The problem is that in a battery a chemical reaction produces the electricity; to charge the battery requires reversing that chemical reaction while maintaining the physical integrity of the electrodes. It is difficult. If the technology improves so that you can recharge the battery 7000 times instead of 700, then electric fuel will cost only 3 times as much as gasoline, instead of 30. There is one additional problem. Gasoline can be pumped into an automobile in 2 or 3 minutes. Charging a battery typically takes 15 minutes to a half hour. Quick-charging batteries are being developed, but they may be even more expensive.

What about the Prius? That's a hybrid—and it uses a battery. And it's a great success! Doesn't that prove that my numbers must be wrong?

Unfortunately, the answer is no. Even the Prius suffers from battery limits, although very few Priuses have been driven far enough to require battery replacement yet. The Prius attempts to minimize the recharge problem by using the gasoline engine most of the time. The battery is recharged primarily during braking, and that helps the car get its good gas mileage. It is also used when accelerating from a stop because that's when the gasoline engine is least efficient. Eventually the batteries will need to be replaced, which may cause the owners consternation when they learn the cost.

Some Prius owners have adapted their cars to "plug in," so that they can recharge their batteries at home, and some add circuits that allow them to drive the car in pure battery mode without the gasoline engine turning on. They say they are saving lots of money on gasoline. That's true. But the more they use the batteries, the sooner they will have to replace them. They may be shocked when they have to pay for the new ones, and I don't mean electrically shocked.

There is one kind of cheap and reliable kind of battery that is not overly expensive to replace: the lead-acid battery used in standard autos to start the engine. This battery is so inexpensive that it can actually be used for an electric car and will cost no more than $3 per gallon of gasoline equivalent even when the replacement cost is considered. The main disadvantage of such batteries is their poor energy storage capability. They typically contain $1/1000$ the energy of gasoline. Combining that fact with the high efficiency of electric motors leads to the conclusion that lead-acid batteries are only 300 times worse than gasoline, pound for pound. That's why they are used for extremely short-range applications, particularly when pollution from carbon dioxide would be a problem. They work effectively for forklifts and wheelchairs but, with the factor of 300, find very little use for normal driving.

Who killed the electric car? Expensive batteries did.

Fusion

Fusion is what powers the sun; it is what powers the hydrogen bomb. The fuel of fusion can be the hydrogen found in ordinary seawater. It is not a "renewable" fuel in the sense that it is used up, but nobody is seriously concerned about running out of seawater. Per pound, the energy is 8 million times greater than that from burning gasoline. Even the energy needed to separate hydrogen from water is negligible compared to the enormous energy that can be released when two hydrogen nuclei fuse.

Humans first "conquered" fusion in 1952 when the US exploded the first hydrogen bomb.[23] They did this by taking two forms of hydrogen, called *heavy hydrogen* (or *deuterium*) and *doubly heavy hydrogen* (called *tritium*), and heating them to many millions of degrees Fahrenheit with a fission-based atomic bomb.

Fusion may have been conquered, but only in the sense that humans conquer a mountain by climbing it. A greater goal is to tame fusion, and this has not yet been done. A fusion reactor must be able to ignite little bits of hydrogen, and to do it using something other than an atomic bomb. Ever since fusion was achieved, it has been a dream of scientists and engineers to utilize this technology for generating nuclear power.

Someday the dream will come true. In the chapter on nukes, I discussed the various technologies being developed. The most prominent ones are the tokamak, a large magnetic bottle designed to contain the hot hydrogen long enough for it to fuse, and laser-induced fusion in little pellets—miniature hydrogen bombs, really—that are ignited by lasers and that control the fusion only by limiting the amount of hydrogen in each pellet.

Many people continue to be optimistic about fusion power, but the history of such optimism leads to pessimism. Over the past 50 years, the prognosis has not changed: fusion power has always been predicted to be viable about 20 years in the future. As the research has developed and problems have been solved, more problems have emerged, all of which have looked solvable. As a result, we are still at a point where controlled thermonuclear fusion, or CTF, may be viable in a few decades. We can't really be sure, so it is not a wise strategy to plan on it. I think the only statement on which there would be a consensus is this: There will be no commercially viable fusion reactors in the next 20 years. That's the key thing that a future president needs to know. Fusion will likely be the energy source of the twenty-second century, to the extent that making predictions like that makes any sense whatsoever.

Solar Power

Many advocates of conservation argue that we should already be adopting solar power on a massive scale. But solar is not likely to solve the problem in the next decade or two, at least not for developing countries that cannot afford to pay the premium. In the chapter on solar power, I discussed the solar thermal plant near Seville, Spain, which works fine but delivers electric power at the relatively high price of 28¢ per kilowatt-hour. To really address the CO_2 problem, of course, the relevant issue is the cost of a similar plant in China or India. That is hard to gauge, since the prices of land, construction, and personnel are all different, as is the availability of cloud-free days. But in China coal is cheaper too. The relative price of solar versus coal may prove to be the most important factor in China's future energy decisions.

What about local solar, installed on the rooftop of your home? I discussed this option in some detail in Chapter 6, Solar Power. I showed there that the cells would have to last, typically, more than 22 years in order to save you money. Right now, solar is an alternative only for wealthy people who wish to reduce their carbon dioxide production. It is too expensive to replace coal as a source of energy for the developing countries of China and India.

When I include solar power in this chapter as a nonsolution, I am referring to current solar technologies. Advanced solar methods do offer improved hope for the future. I'll explain my optimism in Chapter 24, New Technologies.

Recycling

Many people feel that recycling is not a solution to any serious environmental problems, but only to aesthetic ones. Nobody likes to see plastic in the wilderness or in the oceans, and packing straps can strangle a penguin, but recycling is unlikely to have any significant effect on atmospheric carbon dioxide. In fact, in one of the strangest ironies of environmental science, it now turns out that you don't want to recycle newspapers or use biodegradable plastics—at least not as far as global warming is concerned. The process of biodegradation is basically one of bacteria consuming carbon compounds and converting them into carbon dioxide.

We could, in principle, reduce carbon dioxide by planting new trees, but only if we were careful to avoid using them for fuel. Letting them rot also doesn't help; that's just a slow burn. My wife Rosemary, an architect, suggests we use more wood for houses: sequester carbon in your home! But planting trees may be a futile approach. You can gauge the magnitude of the problem from the fact that a single gigawatt electric power plant produces a ton of carbon dioxide every 3 seconds. A thousand such plants (the world energy use is about that) would produce 1000 tons every 3 seconds. A lot of trees would have to be grown to take care of all that CO_2.

The problem seems so daunting that billionaire Richard Branson has offered a $25-million-dollar "Earth Challenge" prize to anyone who comes up with a practical way to remove the CO_2 from the atmosphere. The problem is not only that the tonnage of carbon dioxide is huge, but that the atmosphere is even bigger. The concentration of carbon dioxide is only 380 parts per million, 0.038%. No law of physics says that removal of these large amounts in the face of such enormous dilution is impossible, but I suspect that this prize will not be won within my lifetime, or even during your presidency.

Kyoto

In 1998, then Vice President Al Gore signed a proposed amendment to a treaty called the United Nations Framework Convention on Climate Change (or UNFCCC), in Kyoto, Japan. Ever since, this document has been referred to variously as the *Kyoto Protocol*, *Kyoto treaty*, *Kyoto accord*, or simply *Kyoto*. Senate ratification of the treaty would commit the United States to a reduction in carbon dioxide emissions to 7% below our 1990 level. Because emissions have grown since 1990, the actual cut required works out to about 29% of the levels expected in 2010.

The treaty has been ratified by 164 countries—almost the entire world—but not by the United States, to the embarrassment of many US citizens. In fact, it was not even submitted for ratification by Presidents Clinton or Bush, presumably because they knew it would fail to pass in the Senate.

The main objection to the Kyoto accord is that it places no limits whatsoever on developing countries such as China and India. All of the reductions are supposed to come from the developed world. After all, we caused the problem, right?

Right. We were very likely responsible for at least some of the 1°F temperature rise of the past 50 years. We will very likely *not* be responsible for the predicted 3°F to 12°F temperature rise of the next century, however. The main culprits for that increase will be India and China.

In fact, one US Senate vote has been taken on the Kyoto accord, although not to ratify it. The vote was for the Byrd–Hagel Resolution, and it passed in a very bipartisan 95-to-0 vote. The resolution states that the United States should not ratify Kyoto until the treaty is rewritten to include binding targets and timetables for developing nations.

Why, then, are so many people so enthusiastic about Kyoto? Many Kyoto proponents say that the United States must set an example, in the hopes that one day China and India will follow. Opponents say that China and India are certainly going to follow our example, but not in a way that will help. They will develop their economies as rapidly as possible, just as we did, and then, when their people are as wealthy as ours, they will begin to consider controlling their emissions. Although the carbon dioxide that they now produce exceeds that of the United States, their production per capita is less than one-fourth that of ours.[24] If you were president of China instead of the United States, would you cut back? With a population that still suffers from poverty, malnutrition, poor health, lack of opportunity, widespread illiteracy, and periodic famines, would you slow economic growth in order to keep the temperature from going up a few degrees? Add to this the facts that China has plenty of coal, certainly enough to meet the worst scenarios of the global-warming models, and that it is accelerating its exploitation of that resource.

Another argument in favor of Kyoto is that it set up a mechanism of carbon trading that hopes to bring the free-enterprise approach to solving the carbon dioxide problem. The economically advanced nations that ratified Kyoto are already engaging in such trading, and so are most presidential candidates in the United States. If you can reduce your carbon dioxide emissions below the cap limits, you can sell the credits to other countries or industries that are unable to do so. The idea is that such *carbon trading* (short for *carbon dioxide trading*) will give a financial incentive to develop technologies for efficiency and conservation. Opponents of this system say it allows too much cheating. A collapsing economy in Russia, for example, enabled that country to sell a large number of credits for carbon dioxide that it never would have produced. Trading of credits, in this case, led to an increase in the carbon dioxide dumped into the atmosphere over that which would have been emitted.

Nobody expects Kyoto to be a real solution. As Figure 23.1 shows, the growth of China and India will dominate greenhouse gas increases for the foreseeable future. The role of these countries is made even more stark when you consider the greenhouse gases emitted for a given level of productivity. Which country do you think is least efficient in this measure—that is, produces the most carbon dioxide per dollar of gross domestic product? Which nation comes in second?

Without reading the labels below each bar in Figure 23.2, which shows the CO_2/GDP ratio for several countries, look at the plot and

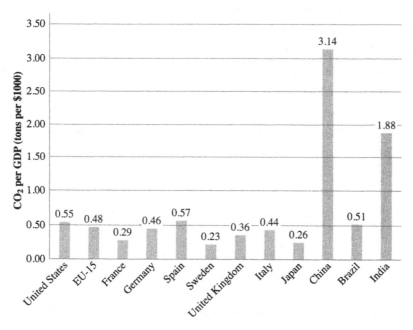

Figure 23.2. Greenhouse gas emission intensity. The ratio of carbon dioxide emissions to gross national product is compared for several countries.

guess which bar represents the United States. Now look at the labels. Are you surprised to learn that the United States is the bar at the left? That the tallest bars (most polluting per dollar) are those of China and India? There is much to be gained, and this plot clearly shows that solutions must be applicable to the developing world if they are to work. This plot also suggests a solution: increased energy efficiency for the developing nations.

In this chapter I focused on things that will not solve the problem. Next we'll talk about approaches that might work.

24

THE FRUIT ON THE GROUND

IN THE previous chapter I dismissed as nonsolutions many of the ideas that are the darlings of environmentalists and elementary school teachers. What is left? A lot. Let's now talk about things that really can work to keep carbon dioxide at reasonable levels.

The first one—conservation—is the most important, the most practical, and the cheapest; in fact, it will earn money. This is the "low-hanging fruit"—the approach that is easiest to take. California Energy Commissioner Arthur Rosenfeld, a physicist, disagrees with the metaphor, however. He says this fruit is not even hanging; it is so low that it is on the ground waiting to be picked up. Conservation is automatic sequestration; it is equivalent to leaving the carbon buried. The trick to conservation—the approach needed to really win over the public—is to do it in such a way that the only way people notice is by the money they save.

Comfortable Conservation

By conservation, I don't mean having a wimpy car that goes slowly uphill. I mean using less energy for identical performance, measured in whatever way the consumer wishes. By conservation, I don't mean turning down your air conditioner on a hot summer day, or wearing a sweater instead of turning up the heat in the middle of winter. I mean setting your thermostat at whatever level you want—but using less energy when you do. By conservation, I don't mean sacrificing your comfort for the good of the world. I mean staying as comfortable as you wish, and saving money in the process. I believe that, to be effective, conservation must be comfortable conservation.

Comfortable conservation is best understood through a historical example: the efficiency of refrigerators. This sounds like a dull subject, but you'll see it is absolutely fascinating, in part because it seems so mundane. Stick with me. It turns out that a little bit of money spent on refrigerators saves a lot of money on coal power plants—and enormously reduces carbon dioxide emissions.

Refrigerators used to use lots of energy to produce meager results. In 1974, the average refrigerator size in the United States had a volume of 18 cubic feet, and the energy it used was 1800 kilowatt-hours per year. That's 130 kilowatt-hours for each cubic foot. It turned out that refrigerators could be made much more efficient by increasing the copper in the motor coils. Of course, that would cost more money—but the cost would be paid for in the reduced electric bills in just the first few months. Think about it: 1800 kilowatt-hours per year costs the consumer $180 annually. Cut the electricity use in half, and you've saved $90 per year—not just the first year, but every year after that. That can pay for a lot of extra copper!

Why hadn't that happened by 1974? Some people think that "market forces" should have pushed refrigerator manufacturers to make the kinds of appliances that the public really wanted. The answer to this conundrum has to do with information. Historically, a consumer in the market for a refrigerator could not tell how much it would cost to operate the appliance; all the sales label stated was the capacity in cubic feet and the price. But then the Federal Trade Commission changed the dynamics of refrigerator purchases by a simple labeling requirement: each refrigerator for sale should include on its price tag a label giving the estimated cost of operation for a year. Suddenly consumers could see that they would save money on slightly more expensive refrigerators—and they started buying them. The manufacturers discovered that the most effective way to sell refrigerators was to make models that would reduce the consumer's electric bills. So they added more copper to the coils and insulated the box itself better.

The results were spectacular. The energy use per refrigerator dropped from 1800 kilowatt-hours per year in 1974, to below 500 at present! Not only that, but during that time the size of the refrigerator increased, from 18 cubic feet to 23. The use of high tech also affected the refrigerator market. The purchase price dropped, when measured in constant dollars, by more than 50%. Win–win!

Does this refrigerator change affect the environment? Yes, it has an enormous impact. If today's refrigerators had the old 1974 efficiency, then the United States would have needed an additional 23 gigawatt power plants (coal? nuclear?) just to supply the added power needed. We've avoided the need for 23 big plants. Way to go! Do you still think refrigerator physics is boring?

Refrigerators provide just one example of the enormous savings that can be gained in comfortable conservation. Everybody won in this example—with the possible exception of OPEC. What more can be done? Lots. Similar gains can be made from efficient air conditioners, which are essentially refrigerators operating on an entire room (dumping the heat outdoors). Perhaps the most dramatic change can come by switching from tungsten lighting to fluorescents.

Compact Fluorescent Lights

Screw-in fluorescent lightbulbs (with pleasant colors) produce the same light as the old tungsten bulbs but use only one-fourth as much electricity (or less). Fluorescent bulbs have a bad reputation, because the older, cheaper bulbs made a light that was not as flattering to human flesh tones as were the redder incandescent bulbs; moreover, they often buzzed or started slowly. But all that has changed, as reflected by the photo of the chandeliers of Notre Dame in Paris that is shown in Figure 24.1. Not only do current screw-in fluorescent bulbs produce a beautiful light, but their lifetimes are much longer than those of the old bulbs. As a result, locations where replacing bulbs is difficult (such as Notre Dame) are rapidly switching over. These new fluorescent bulbs are also highly valued in developing countries, where the cost of electricity is high compared

Figure 24.1. Fluorescent lights in the chandeliers of Notre Dame in Paris.

to income. In my recent trips to Morocco, Kenya, and Rwanda, I saw a higher percentage of compact fluorescent lights than I normally see in the United States (with the exception of my home and that of Energy Commissioner Art Rosenfeld).

A development that may soon be competitive with compact fluorescent lights is the light-emitting diode, or LED. The most primitive of these are the little red lights that tell you if your stereo is on or off. Red LEDs are already being widely used to replace the combination of tungsten lamps and red filters used in traffic lights. White-light LEDs have recently been developed and are already replacing the extremely inefficient bulbs used in flashlights—giving much longer battery life. At present, these LEDs are not quite as efficient as fluorescent bulbs, and they are still more expensive. For that reason, their primary use in the near future will be for specialty applications. The white light they produce has an unpleasant bluish tinge to it, but that could change as the technology improves. They represent, however, a rapidly developing technology that bears watching.

There are many other examples of fruit on the ground. Some people say that buildings in hot climates should all have solar power cells on the roof. It turns out there is a cheaper and more effective solution: simply use "cool roofs," roofs covered with cool colors.

Cool Colors: When Brown Paint Is White

Buildings are warmed, in part, by the sunlight that is absorbed on their roofs. With white roofs, much of the sunlight is reflected. Such roofs can significantly lower the air-conditioning cost. But people don't like white roofs; neighbors complain that they are too bright.

There is a remarkable way to reflect half of the light even with a dark (black or brown) roof. The trick is to use a paint that reflects the infrared radiation but absorbs the visible. It turns out that over half the power in sunlight is in the infrared! So a paint that reflected in the infrared but absorbed in the visible band would look dark to the human eye and yet reflect more than half of the incident power.

Such paints are now being used, particularly in regions of the country that are very hot. Physicists sometimes like to describe such paint with a phrase that they understand but that is very confusing to people who don't understand invisible light. They say that the paint is "white in the infrared." Roofs can be covered with any material that does this. They are called *cool roofs*. Cool paint can be used on cars to reduce the energy spent on air-conditioning and that way

Figure 24.2. Art Rosenfeld.

increase the miles per gallon. Cool colors used on sidewalks, streets, and road pavements reduce the heat island effects in cities.

Reason for Optimism: Rosenfeld's Law

Arthur Rosenfeld (Figure 24.2) is one of the great heroes of comfortable energy conservation. When the 1973 oil embargo hit the United States, he left his research in elementary particle physics behind to begin a program in energy conservation. He founded the Center for Building Science at the Lawrence Berkeley National Laboratory, and that led (among other things) to the development of the compact fluorescent light. In fact, much of what I have described about energy conservation I learned from Rosenfeld and from his colleague David Goldstein. One of Rosenfeld's most remarkable discoveries came from his investigation of the history of energy conservation. What he uncovered gives us not only hope for our future, but also hope for the developing countries.

Rosenfeld compared the energy that it took to produce a dollar of GDP in 1845 to the amount required in 1998. After adjusting for inflation, he found that the amount had dropped by a factor of 4.5. That's an average improvement in efficiency of 1% per year. It has happened naturally, through marketplace pressures. That is important. Along with all other kinds of inflation, our energy use per dollar is going down. That's a reason for optimism.

Can we do better than 1%? Yes, much better. The 1% figure is just an average. During the oil crisis of the 1970s, conservation improved at 4% per year. We currently waste so much energy that Rosenfeld believes that with a little government encouragement, we can sustain a 2% rate per year indefinitely. If you don't want to invest the money, let someone else pay for it. It won't be hard to find someone. Rosenfeld calculates that the yield on conservation investment is 20% per year, tax-free.

Suppose we attain an improvement rate of 2% per year. Compounded over only 55 years, that becomes a factor of three. That means that in 55 years our energy efficiency could be tripled, so we will be using only a third of the energy for the same productivity.[25]

Past conservation efforts have been far more successful than many people appreciate. That's because they have been so comfortable that people didn't notice. Rosenfeld points out that it was conservation that liberated us from the control of the oil cartel in the 1970s. When the oil embargo began, the 4% yearly improvement of conservation in the United States allowed our economy to grow with no increase in energy use. But the members of OPEC, the Organization of the Petroleum Exporting Countries, suffered from their reduced income brought about by the embargo. OPEC was more addicted to US dollars than the United States was to foreign oil. A little bit of extra conservation on our part drove down their income and forced them to increase production. That is a very important point for future presidents to know. OPEC was defeated by US conservation efforts, largely during a period when the general public was unaware that conservation progress was taking place.

Won't all this conservation improvement be canceled by the growth of the world's population? As a matter of fact, no. That's the conclusion reached in the second half of Rosenfeld's analysis.

The Fizzling Population Bomb

I have friends who otherwise like me but consider me morally depraved for thinking that the population bomb is not going to kill us all. It is an unpopular time to be considered an optimist. This prevailing pessimism dates back to 1798, when Thomas Robert Malthus wrote his "Essay on the Principle of Population," one of the most influential treatises ever published. "Population, when unchecked," he said, "increases in a geometrical ratio, and subsistence for man in an arithmetical ratio." In other words, although population doubles every few decades, resources grow incrementally. The dreadful conclusion was that disease and famine were not only inevitable, but served an essential function in reducing population. Some politicians argued that it was immoral to intervene. This bleak outlook gave economics its famous nickname: the "dismal science."

Malthus thought the catastrophe was imminent in 1798. The Earth has had several population doublings since then, but food production has kept pace. Much of the increase in food production came from unexpected developments in fertilization, crop management, and the development of high-yield and insect-resistant crops. Present-day starvation and hunger are caused not by shortages, but by inadequate distribution and inequities in buying power.

Pessimism has not died. In 1968, Paul Ehrlich published *The Population Bomb*, a best seller in which he predicted that mass starvation would devastate the world in the 1970s. Revolutions in agricultural production delayed the crisis, but Ehrlich is still convinced it is coming—and soon. His forecasts have not changed: he has always predicted that the disaster will happen within the next 10 years. So far, he has repeatedly been proven wrong, but he believes that the disaster is inevitable and has only been delayed. Apocalyptic gurus are not deterred by unsuccessful predictions.

Despite the predictions of doom, for the first time since Malthus there is good reason for optimism about population growth.[26] We are in the midst of a remarkable demographic transformation that nobody predicted and we still don't fully understand. The explosive population growth has slowed and will soon turn around. The United Nations now estimates that the population of the world will peak sometime in this century between 9 and 10 billion people. That's higher than the current level—but it is a peak. Starting somewhere in the mid to late twenty-first century, the population will begin a slow decrease. The predictions are now believed to be quite robust, as documented in a careful paper in the scientific journal *Nature*.[27] Malthus's (and Ehrlich's) population bomb is fizzling.

Why? How was our understanding wrong? If each family has three children, then each generation increases numbers by 50% (3 children from 2 parents). After two generations, the population should be $1.5 \times 1.5 = 2.25$ times larger. This seems so obvious. Without disease and starvation, the population rapidly doubles. How can this simple calculation possibly be wrong?

At a United Nations conference in 2002, demographers discussed many possible explanations. Some argued that the declining growth is a consequence of the expanding worldwide rights of women. Another explanation attributed the decline to poverty reduction. Wealthy people have fewer children, for reasons we don't fully understand. Western TV was also cited, because viewers see happy families with small numbers of children. In my recent visits to developing countries (Kenya, Morocco, Paraguay, Rwanda, Costa Rica), I found people determined to limit the number of their children in order to be able to afford to educate the one or two that they would have. One man did not have electricity in his home but felt that hope for the future lay in focusing all his resources on one descendant. It may be that the limited growth of population is a direct result of optimism—the idea that children will survive and can have a better life than their parents had.

Even if we don't understand why, the happiest news comes when we combine limited population with conservation growth. The conservation bomb wins. Rosenfeld points out that at 2% growth—the 2% solution—conservation outruns population by a large factor. Two percent compounded over 100 years reduces energy use by a factor of 7.2. By 2100, with a world population of 10 billion people, everyone can be living at the current European standard of living, yet expending half the energy that we are using today. That includes not only China and India, but also Africa and the rest of what (in 2100) will be the formerly developing world.

What a delightful vision! Economics—no longer the dismal science but the *glorious* science. Wealth reduces population growth; conservation wins; the environment is cleaner; the world is happier.

If we allow conservation to putter along at 1%, on the other hand, then in 2100 we will be using 40% more energy than today.

There is a catch. If the US standard of living continues to increase, and the developing world wants to match that increase, then the energy requirements may continue upward; that might require conservation to reach 3% annual growth.

The 2% solution is painless, but to achieve it requires conscious government-led efforts to develop cleaner, more energy-efficient technologies in areas like power generation, transportation, manufacturing, and environmental control. Cancellation of research programs in these areas is self-defeating. The solution to pollution is conservation. That could be a good sound bite for your presidential campaign.

There are many ways to improve efficiency, now that we have a reason to do that. We've already discussed refrigerators and air conditioners. We'll discuss more examples in the chapter on new technologies. As an easy final example for this chapter, let's take a look at automobiles.

Automobile Efficiency

In physics, it takes no energy to move an object horizontally, except to overcome friction. In some sense that is obvious. You can slide effortlessly over ice. Of course, you put some energy into the object as it speeds up, but you can get that back when you slow it down, provided that you use it to charge a battery rather than to heat brakes. That's what a hybrid auto does.

If it takes no energy to move us, in principle, then why do our autos burn so much gasoline? The answer is that they waste it. They do that in three ways. When you accelerate, you are putting energy into your motion, and when you slow down, that energy all goes into heating the brakes. You lose about a third of the energy that way. Another third is lost because of air friction—pushing you and the auto through resisting air. The force of air resistance increases rapidly with velocity, and it is the reason why some automobiles now get better mileage in the city than on the highway; they go slower. Finally, a third of the energy is lost in heating the road and the tires. All this energy is turned into heat and wasted.

Because essentially *all* the automobile energy is wasted, it should not surprise you that we can do much better by simply wasting less. We can't quite reach the physics limit (no energy to move horizontally), but we can greatly increase the number of miles per gallon. Our present automobiles are, in part, artifacts from the time when gasoline was cheap and nobody worried about pollution. Think back to the refrigerator example in the previous section—and how low efficiency reduced the purchase price, even though it meant higher cost for the consumer in the long run. What could we do to make an auto more efficient? How many miles per gallon could we get?

The first step is to build more and better hybrid autos. They get more miles per gallon by two means. Ordinary cars slow down by converting their energy of motion into heat; that's what brakes do. Hybrid cars slow down (whenever possible) by hooking the wheels to an electric generator. The force needed to run the generator slows the car, but the energy is converted into electricity and stored in the hybrid's battery.

Hybrid autos conserve energy in a second way. Because the wheels can draw energy from the battery, it is not necessary to vary the auto engine speed. That means that they can run the engine at a more constant rate, near the ideal point for efficiency. This effect is particularly important when the car is accelerating from a stop.

Hybrids are not a new idea. Diesel-electric locomotives are hybrids, and they largely replaced steam engines in the early and mid 1900s. Diesel submarines, the kind used in World War II, were hybrids. They charged their batteries when they were on the surface, and then ran silent and deep using lead-acid batteries that filled up the compartment below the submarine floor.

The Toyota Prius, the most famous of the new auto hybrids, gets a nominal 60 miles per gallon. I know nobody who actually gets such high mileage, but when I take my Prius on a long round-trip excursion to a real location and drive carefully, I'm able to coax the car to 50 miles per gallon. But the Prius is a heavy, almost luxury car, undoubtedly designed to help give the hybrid idea high social status. It has several limitations. A pure hybrid (like a diesel locomotive) would run the wheels only off the battery, and would use the gasoline motor only to charge the battery. That way the engine could be kept at its most efficient speed. The Prius does not do this, presumably because its expensive nickel-metal hydride (NiMH) battery can be recharged only a limited number of times. Someday that number will be increased, and the cost of batteries is likely to come down. As we move more and more toward the ideal hybrid design, there is no reason why the mileage could not increase further. In the future, true hybrids may well attain a true 80 to 100 miles per gallon, or even better. Right now, the average automobile in the United States gets about 25 miles per gallon. The US Senate proposed a bill in 2007 that would require auto makers to sell cars with an average fuel consumption of 35 miles per gallon by 2020. This seems like a reasonable approach, but it's the kind to take if you feel there is no crisis. If we felt it were important, we could soon be doing much better. That is something that only a president can decide, because to make it happen the president would have to be the one to convince the public that it was worth doing.

Another key innovation for improving the energy efficiency of cars is to reduce automobile weight. Lighter cars take less energy to accelerate. A perfect hybrid would recoup all that energy and store it in the battery when the car slowed down, but engines aren't 100% efficient, so there are significant gains to be made. A light car also saves energy by reducing the friction on the road, much of which comes from the compression and decompression of the tires as they rotate.

The main objection to light autos is that they are less safe than heavier cars. Many people feel that a heavy car protects them, and they are partially right. In a heavy car, you are more likely to injure people in the car you hit than to be injured yourself. Ironically, if all cars were required to be light, then everyone's safety would be improved. This is an example of a well-known conundrum known as the *paradox of the commons*.[28] If everyone cooperated, then everyone would be better off, but anyone who departed from the common agreement and picked a heavier car would be safer, although the average safety would decrease. Unless a law regulates size, the market will push toward larger cars. The easiest way to have such a law is to prescribe efficiency.

In fact, light cars can be made much safer if constructed properly. We can achieve a lightweight body by making it of fiber composite, a material that is stronger than steel (for the same weight) and is likely safer in a crash because it is better at absorbing energy. The aerospace industry is already converting many of our airplanes to composite materials, and we need to do the same for autos. A process for making and molding such material for automobiles and other needs has now been developed by Amory Lovins, a leading advocate of improved energy use. If the efficiency can be made good enough (Lovins feels that 100 miles per gallon is easy, and 200 miles per gallon is within reach), we could even reconsider hydrogen as a fuel. If all you need to carry is a few gallons to go a few hundred miles, then most of the objections to hydrogen automobiles go away. There is no need to consider hydrogen right now, however. Efficiency in automobiles more than pays for itself. As Lovins likes to say, "Energy efficiency isn't just a free lunch; it's a lunch you are paid to eat."

Hybrid automobiles have gained such favor that they have turned into a marketing ploy. Lexus recently came out with a hybrid luxury automobile that gets only 22 miles per gallon—not much better than their nonhybrid version, which gets 18. The Ford hybrid SUV advertises 34 miles per gallon; that's not bad, for an SUV. Much of the incentive for such cars comes from the fact that hybrids are considered good for the environment, but a hybrid that gets such low mileage does not really qualify. We need incentives to spur the move to autos that truly are more efficient.

I have no doubt that hybrid automobiles will take over the automobile market. Your grandchildren may never drive a hydrogen auto, but they are pretty certain to drive a hybrid. If legal limits are placed on the average gas mileage, we can expect to cut our gasoline use in half while maintaining comparable performance. Of course, it is important to remember than only 28% of the US fossil fuel consumption is used for transportation (including airplanes), so cutting gasoline use in vehicles alone cannot solve the carbon dioxide problem. But no single approach will fix it all, and every little bit helps.

Note that I am not advocating hybrids because of the lowered cost of operation. Until battery costs come down, there may be no net monetary savings. The hybrid is good for society because it reduces fossil fuel consumption and the associated carbon dioxide emissions. Lower oil use gives the United States greater energy independence. But don't plan on buying a hybrid car to save money.

25

New Technologies

FEW would argue with the fact that the first line of defense against carbon dioxide increase must be energy conservation. It is unfortunate that this basic strategy gets very little attention compared to new high-tech methods, but when you are president, you may be able to change that. New technologies are probably the second line of defense, and as such they, too, are important. Because they are high tech, they get most of the press coverage. They also get most of the investment. I won't try to cover the multitude of approaches in this chapter; there are too many. Instead I'll describe a representative sample of the most promising ones.

Biofuels

Grow a plant, burn a plant—and there is no net carbon dioxide increase. That's because plants get all of their carbon from atmospheric carbon dioxide, and burning it just puts that same carbon dioxide back into the atmosphere. For that reason, *biofuel*—fuel made from plants—is said to be *carbon-neutral*. In today's world, biofuels are never completely carbon-neutral, because growing the plants often uses energy and fertilizer made from fossil fuels. As a result, use of corn ethanol reduces net greenhouse gases by only 13% compared to the use of fossil fuel, according to some of the best recent estimates. In contrast, use of ethanol made from sugar cane in Brazil reduces greenhouse emissions by 90%. Almost half of the automotive fuel in Brazil is now sugar-cane ethanol. So we must not lump all biofuels into the same category. Some are good, and some are not so good. Corn ethanol is arguably the worst. Beware: as a presidential candidate, you will be subjected to enormous political pressure from the early-caucus state of Iowa to pledge your support to corn ethanol.

An element of politics and economics works its way into all of these issues. The Brazilian ethanol program was not begun with the goal of reducing carbon dioxide emissions, but rather to enhance Brazil's energy independence during the 1973 oil crisis. The United States is now getting into biofuels in a serious way. In 2005, the United States passed a law mandating the use of 7.5 billion gallons of ethanol fuel per year by 2012. This law was justified by its role in reducing carbon dioxide emissions and in helping make the country energy independent, but there was also a political side. The mandated use of ethanol helped boost corn prices.

One-fifth of the US corn crop is now diverted to making ethanol. According to the *Economist* magazine, demand for corn will exceed supply for the foreseeable future. The price of corn has shot up, land has been diverted from soy and wheat to corn, and that has driven up the price of feed grains, poultry, and meat. This trend is now called *agflation*. Like everything else bad in the economy, agflation hurts the poor most. Rises in the price of tortillas are blamed for food riots in Mexico. Does it really make sense to do all this to save 13% in carbon emissions? This is important for future presidents to know: corn ethanol has little advantage for clean energy, but some alternatives do.

Ethanol is the same alcohol that we have in our wine, beer, and liquor. It has several physics features that affect its use. It can be mixed with gasoline, making a mixture called *gasohol* that can be used in most US autos with little or no adjustment to the engine. Ethanol actually enhances burning of the gasoline in the engine, giving the fuel a higher octane rating than it would otherwise have. Some people like it because it is cheaper per gallon than gasoline, but that is a false economy. Ethanol delivers only two-thirds the energy of gasoline, so although it is usually cheaper per gallon, it is actually more expensive per mile.

Ethanol from Brazilian sugar cane costs only $1 per gallon, much cheaper than US corn-based ethanol. Brazil has lots of land, lots of rain, and an ideal year-round warm climate that allows the growth of several crops each year. In contrast, most of the United States does not have an ideal climate for growing sugar cane. But the Great Plains were once covered with high grass, and that could be made to happen again. In his 2007 State of the Union speech, President Bush specifically mentioned switchgrass, a plant that can grow over 10 feet high in one season. Even better is miscanthus (Figure 25.1), a grass being used in Europe for biofuels. Compare these numbers: an acre of corn can produce 354 gallons of ethanol each year; the same acre growing miscanthus can produce 1150 gallons in a year—over three times more.

There is a technological barrier to using such grasses for autos. Most of the plant material in grass is cellulose rather than sugar or starch. Cellulose can be burned for energy (often mixed with coal, or converted to a synthetic gas fuel called *syngas*), but there is no efficient industrial process to convert that to a liquid fuel. In contrast, sugar and starch (from cane, beets, or corn) can be converted to alcohol in large fermentation tanks. It is considered very likely, however, that we will be able to bioengineer microorganisms to convert cellulose directly to ethanol or other alcohols. The ideal alcohol would be butanol because its energy per gallon is almost equal to that of gasoline.

How much of US farmland would it take to make a difference? Jay Keasling, a scientist at the Lawrence Berkeley National Laboratory, estimates that planting 100 million acres of miscanthus would meet the entire US demand for transportation fuel. That's one-fourth of

Figure 25.1. One of the best crops for biofuels: Miscanthus giganteus, *11.5 feet high.*

the current US cropland. Does that sound like a lot of area? It is a square 400 miles on each side. Too much? Is that the sort of program you might be willing to endorse during your campaign? Or is it too disruptive?

Why bother to convert the grass to ethanol or butanol? Why not just burn the grass and get the heat out directly? That's not a crazy idea; in the early United States, the fuel for the railroads was primarily wood cut from nearby forests. Wood is, of course, the traditional biofuel. Miscanthus is currently being burned for fuel in Europe. The real reason to convert grasses to liquid is the convenience of having liquid fuels—fuels that leave no ash—for use in our autos.

Concentrating Solar

Earlier I listed solar as one of the nonsolutions for global warming, but that's because current solar cells are too expensive for developing countries such as China and India. There is yet the possibility of hope in new technologies; in fact, it is conceivable that solar will eventually become a major source of power. It all depends on price. One of the truly hopeful developments in recent years has been the development of highly efficient solar cells. These are complex devices because extracting as much energy as possible out of sunlight requires having separate layers to convert different colors. These sophisticated solar cells are now being built, and one major producer, Boeing (yes, the airplane company; it started producing solar cells when they were needed for space), will soon be selling solar cells that convert 41% of the incident sun power·to electric power. They say that the efficiency should rise to 45% in the near future. Wow!

There is a catch, of course. Even when purchased in large quantities, these special cells cost about $70 per square inch, about $10,000 per square foot. A foot-sized cell would yield 41 watts—not much for the $10,000 investment. Why do I call this hopeful? The reason is that sunlight can be focused using a lens or mirror. You can make a plastic lens that is 1 foot square for less than $1 and use it to focus the sunlight onto a cell 0.4 inch on a side. A cell of that size costs $10. Your total cost for the 41 watts is now reduced to $10, plus $1 for the lens, plus whatever you spend to build the module. That sounds *very* attractive. The tricky part is that you have to keep the cell pointed at the sun, and that requires a mechanical system. If our goal is to spend no more than $1 per installed watt, then the total cost for the square-foot device must be less than $41. Can that be done? It is not obviously impossible, and several companies in California are already building such systems to see if they can be cost-effective. Even if it costs three times that, this system still becomes the cheapest form of solar power.

This approach is called *solar concentrator technology*. Its greatest drawback is that it works only on sunny days, when the sun is visible and its rays can be focused. In contrast, nonfocusing systems can deliver substantial power even when it is cloudy, because they can still work with the diffuse (but dimmer) light that reaches the ground after being scattered by clouds.

Imagine now an array of foot-sized concentrator solar cells covering a square mile of sunny Nevada. Since there are 5280 feet in a mile, there would be 5280 × 5280 = 27,878,400 modules. Each module would be only a foot high, making the system quite robust against wind. Driven by tiny electric motors, the modules would all point in the same direction: toward the sun. With 41 watts from each, the total electric power output at midday would be over a gigawatt.

Another promising technology is the ultra-cheap solar cell called CIGS (Copper Indium Gallium Selenium—the elements from which it is made). CIGS cells have reached 19% efficiency, and they may soon cost less than the canonical $1 per installed watt. They are not made of crystals, but are evaporated or sputtered on glass or film. Watch for new developments; investors are so optimistic that factories costing over $100 million are under construction.

Safe Nukes

In the late twentieth century, a moratorium on nuclear power went into effect in the United States; existing plants continued, but no new ones were constructed, ostensibly because it was considered unsafe. In fact, however, its safety was never explicitly compared to that of fossil fuel. The top dangers that people cited at the time were

- *Meltdown*, the danger of a catastrophic accident
- *Diversion*, the danger of a "plutonium economy" with nukes everywhere
- *Waste*, because nuclear fuel radioactivity lasts thousands of years

Some people argued, even back then, that the fear of nuclear power was exaggerated. Notable among them was Bernard Cohen,[29] who wrote a large number of articles and books with calculations showing that the dangers were actually quite small compared to many others that we find acceptable. That was before the dangers of global warming were added to the risks of fossil fuel. These days, many would argue that risks of fossil fuels should also include the catastrophe of war.

There appears to be a growing sentiment that nuclear power should be reconsidered. Several technological achievements have encouraged this feeling. Significant engineering and scientific developments have made this technology even safer than in the past. One outstanding example is the pebble bed nuclear reactor, described in Chapter 12, Nuclear Power. Such a reactor was built and operated in Germany, although it was shut down to quell public protest after a small release of radioactivity. Pebble bed technology is now actively being developed at MIT and by companies in South Africa, the Netherlands, and China. It has a major advantage for future presidents: the pebble bed approach is considerably safer than the old nuclear reactor design. Many analysts, including those who previously opposed nuclear power, now regard the dangers in this new design as considerably less than those of fossil fuels. Because the fuel pellets are difficult to reprocess, pebble bed reactors are also safer against the dangers of plutonium diversion. The disadvantage is that the plutonium cannot be easily removed from the pellets for use as a fuel itself.

Are pebble bed nukes the answer to all our energy problems? They could supply much of our electricity, but they don't provide the kind of transportable energy that is needed in our autos. Moreover, the supply of minable uranium is limited, and to continue into the long-term future, we might have to either use breeder reactors or mine uranium from seawater. (Seawater contains only 3 parts per billion of uranium, but there is a lot of seawater; separation techniques are being pursued.) A good argument can also be made that present-day reactors are safe enough that we don't even need advanced designs. Pebble bed reactors are also up against the exaggerated public fear of radioactivity. Future presidents will have to address that issue. But the real challenge for pebble bed reactors, or any kind of new technology, is that it must face the competition of cheap coal.

Clean Coal

In the United States, coal retails at $40 to $80 per ton. In Australia, it costs only $2.50 to mine that much. Think about that. For equal energy, you'd require over 200 gallons of gasoline. At $3.50 per gallon, that's $700. Compared to gasoline, coal is dirt cheap.

Coal is not only cheap but abundant. Moreover, the countries that are expected to need the most energy in the future—China, India, the United States, and Russia—have lots of coal, enough to last for centuries. That is good news for the economy, and maybe also for world peace; if we all start using coal instead of oil, these big countries will have fewer conflicts over energy supplies. Moreover, coal can be converted to liquid fuel, convenient for automobiles, through the Fischer–Tropsch process.

The bad news is that coal produces more carbon dioxide than does either oil or natural gas for the same energy. That's because coal is primarily carbon, and all that carbon is turned into carbon dioxide. Oil and natural gas, in contrast, are hydrocarbons, and substantial energy in these comes from the burning of the hydrogen to make the relatively innocuous compound dihydrogen oxide, otherwise known as water. So not only is coal dirt-cheap, but it is also dirty.

Clean coal is possible, however. When the coal is consumed in a centralized plant, the carbon dioxide can be captured and sequestered underground. The process sometimes goes by the acronym CCS— standing for *carbon capture and storage*. It is a big, elaborate, and expensive approach, but it will probably work. Sequestration is not a new idea. A million tons of carbon dioxide is currently being pumped into the ground each year into wells in the North Sea. Sequestration is also being used in Canada. Oil companies presently pump CO_2 down into the ground as a way of forcing up additional oil out of the partially depleted oil reservoirs. Sequestration is also being used in Norway's Sleipner gas field, and at the Salah plant in Algeria.

The IPCC, the organization that has made the most detailed and careful estimates of carbon dioxide effects, has also studied sequestering. Excellent locations for long-term storage are oil and gas fields, coal beds, and underground brines. These salty aquifers are very common, are found all over the world, and were previously thought to have no economic value. The CO_2 dissolves readily in the water, and the IPCC estimates that the gas will remain in these deep locations for millions of years. Another possible location is the deep ocean, although this is not very convenient for most power plants. The IPCC estimates that in such storage location sites, 99% of the gas will remain stored for over 1000 years. The CO_2 would be injected at great depth (up to 2 miles), although some people worry that the resulting local acidification of the water could harm indigenous species. The cost to sequester in this way has been estimated at $40 to $80 per ton—comparable to the cost of coal in the United States. So the price of coal would go up, but it would still remain much cheaper than gasoline.

The United States had recently planned a highly efficient 275-megawatt power plant called *FutureGen*, designed to be the first fully CCS power plant in the world. Proponents hoped it would prove the feasibility of clean coal. It used an elaborate new design called the *IGCC*. (The letters stand for *Integrated Gasification Combined Cycle*—but you don't need to remember that.) Small IGCC demonstration plants have been running at two sites in the United States and two in Europe; larger IGCC plants are under development in the Netherlands, Spain, and Italy. But in early 2008, the United States abruptly cancelled the FutureGen program. Was that the end of the IGCC in the United States? Not necessarily. A demonstration of the technology may have been premature. One expert told me that in its rush to completion, FutureGen would only demonstrate the high expense of IGCC, and make people believe it impractical. A slower but more thoughtful development could result in lower costs and lead to a greater future for this technology.

In an IGCC plant, air is first separated into oxygen and nitrogen. This extraction takes energy, but it allows the plant to burn pure oxygen, and that makes the plant hotter and thus more efficient. Coal, oxygen, and water are then all reacted together. The result is hydrogen gas and carbon monoxide (CO), a combination that has traditionally been called *syngas* (for *synthetic gas*, as opposed to natural gas). That's why the G appears in IGCC—because the coal is "gasified." It also takes energy to produce syngas, but the plant will get that energy back when it burns the mixture. In this gas form, it is relatively easy to remove some of the more noxious pollutants from the coal, including mercury and sulfur.

In the IGCC, the syngas is then burned, producing energy, CO_2, and water vapor. The energy is used to run a turbine. The gas that comes out of the turbine is still pretty hot, and this waste heat is not completely wasted but used to heat water to run a second turbine. So there are two turbines; that's why it is called a *combined cycle* system. Some of the energy produced must be fed back to separate the oxygen from air. Even so, the IGCC is expected to turn as much as 45% to 50% of the coal energy into electricity (those are the goals of the engineers, not yet proven)—considerably higher than the 35% that can be obtained from an ordinary coal-burning plant.

An IGCC plant certainly can be built. The only question is whether it can be built cheap enough that we can afford to subsidize it to make it competitive with dirty coal plants.

The main disadvantage of sequestration is the cost. The IPCC report in 2005 estimated that sequestration would increase energy costs by 30% to 60%. The United States could afford such an increase, but could China and India? Should they bear that cost, at the expense of the other needs of their people? The solution might be to subsidize their costs through carbon credits.

Carbon Credits

Given the complexity of the problems with carbon dioxide emissions, what is the best approach to a solution? Solar plants cost more to construct than coal plants, but they don't pollute. Economists argue that the cost of coal is actually higher when you consider the extrinsic costs—the ones that don't appear in your monthly bill. In this point of view, if we included the damage done by CO_2 in the cost of coal, then solar might actually be cheaper. One suggestion for solving this conundrum is to include the cost of pollution in the price of doing business. The means to this end could be a pollution tax, but many people believe a better solution is a pollution market-place. In such a scenario, if you were going to pollute, you would have to buy pollution credits—more pleasantly called *carbon credits.* A carbon credit could be sold by someone who could conveniently reduce carbon dioxide emissions, to someone for whom it would be expensive to do so. The credit market would be arranged to reduce the total CO_2 emissions. If the market worked, then the net effect would be to set the cost of CO_2 pollution at the value representing its true cost to society.

This is, of course, economic theory and not physics, so I won't try to analyze it here. Carbon trading has been officially established for the countries who have ratified the Kyoto Protocol, and the current price is about $30 per ton of CO_2 emitted. Informal carbon credits have been placed on the market by people in the United States who want to partake but live in a country that has not ratified Kyoto. Several presidential candidates, fearful that they are setting a poor example by flying around the country in private jets that produce lots of CO_2, have publicly announced that they are offsetting the harm by buying carbon credits, so that the net effect is good. Others complain that this is just a system to let the wealthy pollute. Of course, the wealthy will always have more money than the poor, so some people counter that this is an appropriate way both to reduce CO_2 and to transfer some of the wealth from the rich to the poor.

If the carbon-trading market remains at about $30 per ton, that is a good message for many clean technologies. Estimates for the price of sequestering, for example, range from 50¢ to $8 per ton. You can make money by sequestering, provided that the price of CO_2 credits stays high. There is, of course, a danger that the price is artificially high because of limited means to reduce emissions; there are many buyers but few sellers. One must be careful before investing in such a market.

Carbon trading will not necessarily reduce CO_2 emissions. Whenever China builds a new coal plant, it can earn carbon emission credits by making the plant cleaner than it otherwise might have been. That means that China earns credits while increasing pollution. Even so, carbon credits may prove their value by providing a politically acceptable way to transfer money from the developed countries to the developing ones. In effect, the developed nations would be subsidizing the added expenses of clean energy in the developing world. This may be the only way to prevent carbon dioxide emissions from escalating out of control.

Renewable Energy

What word dominates the public discussion of clean energy but has been mentioned only once in this book so far? The answer is *renewable*. Many people seem to think that *renewable* means "good." That's not an equivalence born in physics, but is merely a shorthand for people to divide mostly good from mostly bad. It is important for a future president to know the details, however.

Technologies generally listed under the title *renewable* include solar, wind, biofuels, hydroelectric, and geothermal. They are not all environmentally benign, however. Hydropower is classified as renewable, but many environmentalists oppose the building of new dams. Some even seek to remove existing dams, such as the O'Shaughnessy Dam in Yosemite National Park, which has flooded the beautiful Hetch Hetchy valley.

Sources of energy that are usually labeled as nonrenewable include all fossil fuels, nuclear fission, and nuclear fusion. *Renewable* is an unfortunate word choice because it does not reflect the carbon dioxide problem. Conservation is the easiest and cheapest way to reduce carbon emissions, but it is not listed on the Department of Energy Web site as a renewable source, and perhaps for that reason it doesn't get the attention it deserves. Nonrenewable sources that emit no carbon dioxide include fission and fusion. Given the fact that fusion would use water as its fuel source, it seems meaningless to complain that it is not renewable.

Geothermal energy uses the heat that comes from the ground. The ultimate source for this heat happens to be nuclear: the uranium, potassium, and thorium found in the crust and mantle of the Earth are undergoing radioactive decay, and the heat from these decays is what powers our geysers, volcanoes, and general heat flow from the ground. The average heat flow from the Earth is about 0.007 watt per square foot. Compare that to the 100 watts per square foot available with peak sunlight; it is weaker by a factor of 14,000. For this reason, geothermal is practical only in locations where the Earth tends to concentrate it naturally, such as areas of hot springs, geysers, and volcanoes. Elsewhere the energy is dispersed and difficult to concentrate, since the thermal conductivity of the Earth is low. Geothermal energy is not likely to be a major alternative to other kinds of energy. Optimistic statements about geothermal energy usually ignore the huge cost involved in extracting anything beyond the easiest sources.

Another renewable resource—wind—has been romanticized in the enchanting windmills of history. The windmill was originally a mill (a factory for grinding flour) driven by wind power, although early windmills were also used by the Dutch for pumping water out from behind their dikes. Modern windmills are not as charming as their historical counterparts. They have been optimized by scientists and engineers to extract as much power from the wind as possible, and the results are fantastic. Because it is important to collect not only surface wind, but also wind from higher altitudes, the modern windmill is taller than the Statue of Liberty. These are more commonly called *wind turbines* (Figure 25.2), since they no longer mill flour.

Windy places have been used as sources of power for nearly 1000 years. Because differences in temperature are what drive the winds, wind power derives ultimately from solar energy, which heats the surface of the Earth unevenly. Uneven heating has many causes, including differences in absorption, differences in evaporation, or differences in cloud cover. Windmills must not be spaced too closely, because when a windmill takes energy from the wind, the wind velocity is decreased and the wind is made turbulent; that is, it no longer flows in a smooth pattern. Pilot wind generation plants using wind turbines like those shown in Figure 25.2 were installed at the Altamont Pass in California in the 1970s and are still operational.

Figure 25.2. Wind turbines at the Altamont Pass in California.

There is significant potential for wind power in several regions of the world, including the Great Plains in the United States and certain regions of China. According to an estimate by Nate Lewis, an energy expert at Caltech, we could obtain practical power at the level of 2000 gigawatts around the globe using wind power; that's about 15% of the current world energy use.

A "forest" of wind turbines has been proposed for construction on the ocean off the coast of Massachusetts to supply commercial power (Figure 25.3). This wind park would consist of 170 large windmills situated in a square measuring 5 miles by 5 miles and connected to land via an undersea cable. Each windmill would rise 426 feet from water level to the tip of the highest blade (the height of a 40-story building). The windmills would be spaced ½ mile apart. The maximum power that this forest would be able to deliver would be 0.42 gigawatt. The major opposition to the idea appears to be coming from environmentalists who argue that the windmill array would destroy a wilderness area, would kill birds, and would create noise that could disturb marine animals. I repeat, renewable energy is not always environmentally friendly.

Figure 25.3. Proposed location of a wind turbine park off the coast of Massachusetts.

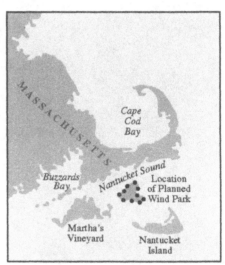

To avoid the problems with the concept of *renewable*, some people have begun to substitute the term *sustainable*. Environmentalists are presently engaged in a debate over what energy sources should be characterized by this attractive word. Is coal sustainable, since it lasts for centuries, even though it pumps carbon dioxide pollutant into the atmosphere? Most people say no. What about nuclear power plants? They don't emit CO_2, but we will run out of the easily mined uranium this century; on the other hand, uranium in the sea will certainly last for millennia. But are vast amounts of nuclear waste sustainable? The debate is not just semantics, since classification of sources into attractive categories provides a simple way for the public to make up their minds on which approaches are good and which are bad. Presidents, however, need to have a deeper appreciation of the subtleties of the issues.

GLOBAL WARMING:
PRESIDENTIAL SUMMARY—
THE BUCK STOPS HERE

AS PRESIDENT, you may be confronted with war, terrorism, oil shortages, and many other threats to US national security, but the most complex physics issue you will face may well be global warming. Unfortunately, you will not be able to master the complex science of the atmosphere, the dangers of carbon dioxide, and the unknown but potentially dangerous positive feedback effects. Almost nobody has. In this book I have just touched on the subject, and I had to leave a great deal out. What is a president to do?

The subject is not made easier by the abundant exaggerations on both sides. Some say that warming will be good because it will enhance crop production and take the chill out of the air. Others show images of Florida and Manhattan under water and suggest that, unless we act, we should expect countless cities to be destroyed like New Orleans. Hyperbole is rampant, and it doesn't make your job any easier.

Let me try to summarize what we know and don't know. Humans are responsible for a 36% increase in atmospheric carbon dioxide, most of it in the last century. Soon, possibly within your presidency, the increase will probably surpass 100%; that is, the carbon dioxide in the atmosphere will be double its historic level—a level it has not exceeded for 20 million years. Huge and complex computer models are working to figure out what such a level will do to the biosphere. These programs are hopelessly inadequate and easily attacked as such. They predict that temperatures will rise between 3°F and 10°F in the coming decades. Such an increase could be extremely disruptive to world affairs, not only because of the temperature change itself, but also because of associated changes in climate patterns. Maybe hurricanes will increase; maybe they won't. Maybe rainfall patterns will change; maybe they won't. Maybe the region of the

North Pole will lose most of its ice; maybe it won't. It seems likely that the ocean surface waters will become significantly more acidic. That change may be disastrous for ocean life, or maybe not.

The danger in exaggerating our certainty is that we could actually have a period of cool weather in the next few years. The hottest year on record was 1998; we have not exceeded that temperature in a decade, and the average temperature over the next 5 years might also remain low. Climate has a variable and random component, and the alarmists might lose their acolytes if additional warming doesn't continue in the near term. But even if that happens, real dangers persist. Our current 36% increase in carbon dioxide is barely having discernible effects (despite the exaggerations of the most vocal alarmists), but when the increase reaches 100% and beyond, the effects should be more obvious. The real danger in shouting that the sky is falling is that it might not fall right away and people will lose interest.

All presidents have to cope with uncertainty. That's what makes the job so tough. You can't afford to be optimistic; the downside is too awful. Yet there are real dangers in being overly pessimistic; the expense of coping with the carbon dioxide increases can be huge. Remember this: the real reason we are worried is not that complex computer models predict dire consequences. The reason is that simple physics calculations of the greenhouse effect show that we are entering an unprecedented regime of climate uncertainty in which human emissions *could* soon cause disruptive changes. Does a president dare take the risk of not addressing these concerns?

Certain steps are pretty obvious. Conservation can be comfortable and can actually save money. It has value for other national issues, such as energy independence. It has been neglected in the past by the federal government, and deserves central attention.

Ultimately, the real issue is the developing world. Freezing carbon dioxide emissions in the United States—even cutting back below 1990 levels—delays the effects of carbon dioxide by only a few years. The booming economies of China and India will quickly fill in for the United States. If you address global warming only with feel-good measures (such as encouraging the use of expensive solar power in the United States), then you are avoiding the real issue and not fulfilling your responsibility as president.

What can you do? If I were president (fortunately there is no chance of that happening), I would fund strong programs in energy efficiency and conservation. I would raise the miles-per-gallon of autos substantially by increasing the government CAFE (Corporate Average Fuel Economy) requirements. I would support efforts at clean coal, particularly efforts at sequestering carbon dioxide. I would encourage the development of nuclear power, particularly fission plants (because I think fusion is too far in the future to really help). I would work to convince the public that nuclear waste storage is a solved problem, and make sure that Yucca Mountain is readily approved. I would see that China and India are given substantial carbon credits for building IGCC coal power stations and nuclear plants. I would encourage technical developments in solar and wind technologies. I would cancel subsidies on corn ethanol and put all the biofuel emphasis on switchgrass, miscanthus, and other more efficient crops. I would encourage developments in efficient lighting, particularly fluorescent lights and light-emitting diodes. Insulation is better than heating; cool roofs (not necessarily white) save more energy than air conditioners and are often a better alternative to rooftop solar cells. I would keep in mind that my emphasis must be on technologies that the developing world can afford.

The really good news is that we are currently wasting vast amounts of energy. That is even truer in the developing world. Conservation is the greatest investment. After all, the cheapest way to remove carbon from the atmosphere is never to put it there. Sequester it in its current form; keep coal, oil, and gas underground.

You have the toughest possible job ahead of you. The buck will stop with you. You've taken the first step in learning some physics. Keep it up. Learn more. Best of luck.

Notes

I ▪ Terrorism

1. I'll put some calculations in footnotes. You are welcome to skip these, since they are not critical for future presidents to be able to do. The hardest (and most boring) part is to get the units right. I will round off all results, since we are interested only in approximate numbers. In the equation for kinetic energy, $E = \frac{1}{2}mv^2$, we must use joules for energy (E), kilograms for mass (m), and meters per second for velocity (v). For this case, the mass of the plane, 131 metric tons, is first converted to 131,000 kilograms. The velocity 600 miles per hour is converted to 270 meters per second. The kinetic energy is then $\frac{1}{2}mv^2 = 4.7 \times 10^9$ joules $= 1 \times 10^6$ food calories $= 1 \times 10^6$ grams of TNT equivalent $= 1$ ton TNT equivalent.

2. In this calculation, I am using a physics principle called the *conservation of energy*. For this situation, the principle says that the force f applied to the hammer multiplied by the distance D over which that force is applied (the length of the hammer swing) is equal to the energy of the hammer. That energy all goes into the nail and is equal to the force F on the nail multiplied by the distance d that the nail moves. In equation form, $f \times D = F \times d$. That makes the force on the nail $F = f \times (D/d)$.

3. This section is based largely on my *Technology Review* essay "The Dirty Bomb Distraction" (June 23, 2004, http://muller.lbl.gov/TRessays/29-Dirty_Bombs.htm.

4. Al-Qaeda's attack may have been *too* successful. It was so devastating that the US public was willing, at least for a time, to support invasions of both

Afghanistan and Iraq. If al-Qaeda's goal was to get the United States to withdraw from the region, a smaller attack might have been more effective.

5. Bryant's story had been disputed, in part because of her claim that Atta visited her in early 2000 when he was reportedly not in the United States. Later evidence supports the possibility that Atta was indeed in the United States at that time, so the criticism is no longer valid. The original interview between Bryant and ABC News can be read at www.muller.lbl.gov/pages/Atta-Bryant.htm.

6. The report was withdrawn soon after the US anthrax attacks but now has been reposted at http://stinet.dtic.mil/oai/oai?&verb=getRecord&metadataPrefix=html&identifier=ADA399955.

II • ENERGY

1. Power is the energy delivered per second; it is a rate of energy delivery.

2. Electric energy can be used more efficiently than the heat energy from gasoline, at least for turning the wheels of a car, and that increased efficiency reduces the disadvantage of batteries by about a factor of three. Thus, gasoline actually outcompetes expensive computer batteries for automotive energy by a factor of only 30. We'll return to this when we discuss electric cars in more detail.

3. This footnote is meant for people who have taken high school physics, even if they have forgotten it, and want to see the calculation. Assume that the bullet moves at the speed of sound, about 340 meters per second. (Some bullets go slower, some faster; but the 340 figure is typical.) The energy of a 20-gram (0.02-kilogram) bullet is given by the kinetic-energy equation (something that future presidents *don't* need to know): $E = \frac{1}{2}mv^2 = \frac{1}{2} \times 0.02 \times 340^2 = 1156$ joules. Gasoline has 10 food calories per gram = 41,800 joules/gram, or 836,000 joules for 20 grams. That's 720 times bigger than the bullet's energy. A Mach-2 bullet would have four times that energy.

4. Of course, exercise has many other health benefits. It's just a difficult way to take off pounds, except to the extent that it reduces appetite. It's often been said that the most important exercise for losing weight is the act of pushing the food away.

5. A future president would not need to know the conversion factor, but here it is: 1 food calorie is approximately 4184 joules. The joule was named after James Joule, a physicist who studied heat and its relationship to energy.

6. One horsepower is actually 746 watts, not 1000, but for most of the needs of the presidency, the approximate value is all you need—and it is easier to remember.

7. The more accurate number is that 1 watt-hour is equal to 0.86 food calorie.

8. I have deliberately rounded these numbers to make them easier to remember. An energy expert needs to know the exact figures; for a president, approximate values are fine. If you want a more detailed breakdown, see the energy flow chart published by the Lawrence Livermore National Laboratory at https://eed.llnl.gov/flow/02flow.php.

9. 24 hours per day × 365 days = 8760 hours in a year. (I ignore the correction for leap years.)

10. We set $14,000 equal to the present value of an annuity of 3%. Then the number of payments of $876 to be worth this present value is 22. This calculation can be done on a spreadsheet using the PV function.

11. To call it 100% efficient is a bit misleading. If you use solar power to generate electricity (at, say, 15% efficiency), and then use that electricity to run a heat pump (with a coefficient of performance higher than 6.7), in principle you can deliver heat at a greater rate than direct sunlight can. In other words, you can have a heat transfer efficiency of greater than 100%. This trick is possible because you are transferring heat rather than generating it.

12. Triple junction cells are manufactured by Boeing-Spectrolab. The 41% efficiency has been certified by the US Department of Energy.

13. The limiting material may be indium, one of the elements used. Most of the world's indium is now used to make thin LCDs (liquid crystal displays) for TVs and computers. Worldwide production is only a few hundred tons per year.

14. The people of Seville use, on average, only 300 watts per person, about one-third the power per capita used by the US population.

15. Hubbert's paper, "Nuclear Energy and the Fossil Fuels," is available at www.hubbertpeak.com/hubbert/1956/1956.pdf.

16. From the Association for the Study of Peak Oil and Gas (ASPO).

17. The leftover oxygen from the water becomes carbon dioxide.

18. Adapted from www.chartoftheday.com.

III · NUKES

1. The energy of these particles is often measured in million electron-volts (MeV). That is the energy that the particle would have if it were emitted from a million-volt generator. A 1-MeV electron has a velocity equal to

94% that of light. A gamma ray *always* comes out with the velocity of light because actually it is nothing more than a very high energy light particle.

2. The rem is the key unit, but sometimes people quote others. One *millirem* is a thousandth of a rem. X-ray a tooth, and it gets about 1 millirem. Other terms you'll see in the newspapers are *sievert* and *gray*. For our purposes, a sievert is equal to a gray, and they both equal about 100 rem. Don't worry too much about learning all of these units. Stick to the rem, and look up the conversion here if you ever need it.

3. Well known among physicists is the fact that 1 rem results from the exposure of every square centimeter of your body to 2 billion gamma rays of 1 million electron-volts each (a typical energy). I take a typical projected area for a human to be 6 square feet, and that combination gives the result of 10 trillion gamma rays.

4. Multiply 2 by itself 10 times and you get approximately 1000: $2 \times 2 \times 2 \times 2 \times 2 \times 2 \times 2 \times 2 \times 2 \times 2 = 1024$.

5. *Ionizing* refers to fact that the radiation knocks electrons off the atoms near its path as it passes by. It does this because the radiation has electric charge which puts a force on the electrons.

6. This calculation is illustrative but circular, since cancer among survivors of these attacks played a major role in deriving the 2500-rem cancer dose number.

7. You could also get this number by dividing 20 rem by the 2500-rem cancer dose: $20/2500 = 0.008 = 0.8\%$.

8. Later in this chapter I'll describe the half-life rule. These numbers don't follow this rule, because they are the result of a mixture of many materials with different half-lives. The short-lived materials decay first, and the long-lived ones last.

9. Suppose, for the sake of argument, that the numbers are the same for mice as for men. Then 20% of the million mice would die of "natural" cancer. That's 200,000 mice. According to laws of statistics, the uncertainty in this calculation would be given by the square root of 200,000, which is 450 mice. Let's now compare that to the expected number of excess cancers. From the linear hypothesis, the risk to each mouse would be $1/2500$. For a million mice, that means 400 extra cancers. But that is less than the statistical uncertainty! To do the experiment properly, you would need to test at least 10 million mice. Of course, even if you manage to carry out one of these "megamouse experiments," there is no guarantee that mice and men are similar in their response.

10. ½ multiplied by itself 10 times is $(1/2)^{10} = 1/1024$. There is no need for a

future president to know the number this accurately, so I approximate it as $\frac{1}{1000} = 0.001 = 0.1\%$.

11. A gram of material with a half-life of 30 years will have half of its nuclei explode in that period. A gram of material with a half-life of 300 years will have only one-tenth as much radioactivity, but it will have it 10 times longer. For a half-life of 3000 years, the rate of decay is $\frac{1}{100}$. Your body will be affected only by those decays that take place during your life.

12. I am not worried about the radiation in my watch, because the electron that tritium emits doesn't have sufficient energy to escape the watch case. It's not just the number of decays per second that counts, but their ability to enter my body and do damage. Beware of radium watches, however; their gamma rays easily exit the watch case.

13. This name does not derive from the word *radio* (which is itself short for *radiowave receiver*). Rather, it is a short form of *radioactive carbon*.

14. Plants absorb carbon-14 from the atmosphere during photosynthesis. The atmosphere, in turn, gets its radioactivity from cosmic radiation (originating in space) hitting nitrogen atoms.

15. Carbon in living material has 12 decays per minute per gram of carbon. Carbon is 18% of the human body. For a 56-kilogram (123-pound) person, that's about 10,000 grams of carbon.

16. One way around this limit is to count the remaining carbon-14 atoms, rather than the number of decays. Even when the decay rate is low, a very large number of atoms remain. The trick is to find the atoms, but this can be done with an instrument known as an accelerator mass spectrometer (AMS). The first person to suggest this method and succeed in doing such a measurement was the author of this book.

17. There is no physical or biological reason to consider such alcohol more dangerous. The rule was probably put in place to keep drinking alcohol expensive. That reduces consumption, and helps some farmers.

18. At an EPA Web site (www.epa.gov/radiation/students/calculate.html), you can calculate the radiation dose you would get in different locations around the country. To convert the answer from millirem to rem, divide by 1000.

19. UNSCEAR's report is available at www.unscear.org/docs/reports/annexj.pdf.

20. For an interesting article about these machines, see David R. Lapp, "The X-Ray Shoe Fitter," *Physics Teacher* 42 (September 2004), 354–358.

21. 50 rem to my feet is equal to an average of 2.5 rem to my whole body. To get the linear hypothesis cancer probability, divide by 2500. That gives $\frac{2.5}{2500} = 0.001 = 0.1\%$. Add that to the nominal risk of 20% that I would have had anyway.

22. Superman would also not be able to look through uranium or plutonium, but I don't think that fact ever made it into the stories. For some unexplained reason, he was highly allergic to kryptonite, which I presume is a compound containing the element krypton. Maybe that sensitivity came from too much childhood exposure.

23. These numbers work out simply because a molecule of TNT weighs about the same as an atom of U-235.

24. This value can be calculated from numbers given in Serber's book *The Los Alamos Primer*. Serber assumed that the bomb released 15 kilotons of TNT equivalent, and he states that only 2% of the uranium actually fissioned.

25. Many books continue to state, incorrectly, that the blast was 20 kilotons. That's the number that President Truman announced, but he mistakenly cited the number for the Alamogordo test, not Hiroshima. My mentor, Luis Alvarez, was flying over Hiroshima when the explosion took place. His job was to measure the energy released. The number he obtained was 13 ± 2 kilotons TNT equivalent.

26. The number refers to the weight of the atom, or *atomic weight*, which is the total number of protons and neutrons in the nucleus. U-235 has 235 neutrons and protons; U-238, the more common kind of uranium, has 238.

27. The term *critical mass* has worked its way from physics into our everyday language as a metaphor. One or two people, working on a problem, may not be enough to solve it. But if you assemble a critical mass of people, the progress (in this metaphor) can be explosive.

28. This value, equal to 200 kilograms, is described in Robert Serber's book, *The Los Alamos Primer*, p. 27. It was based on a diffusion-theory calculation.

29. The nucleus of uranium-235 has 92 protons and 143 neutrons, making a total of 235. The nucleus of uranium-238 has 92 protons (just as U-235 has; that's why it is still called uranium), but 146 neutrons, three more than in uranium-235. Those three extra neutrons tend to bind the nucleus more tightly, preventing the chain reaction. Atoms with different numbers of neutrons are called *isotopes* of the element.

30. In elementary physics, some people teach that the centrifugal force doesn't really exist. But that is true only for inertial reference frames. In a rotating frame (used in advanced physics courses) the centrifugal force is as real as the gravitational force.

31. Deuterium is different from ordinary hydrogen in that its nucleus contains an extra neutron. The tritium nucleus contains two extra neutrons.

32. A simple way to see this is to bear in mind that the volume of the blast, roughly proportional to the yield, is the cube of the radius. If the yield goes

up a factor of 50, the radius goes up by the cube root of 50, about 3.7.

33. This calculation is again based on the cube root rule, explained in note 32. The cube root of 1/8 is 1/2. (That's because 1/2 × 1/2 × 1/2 = 1/8.)

34. For up-to-date information, I recommend the Web site maintained by the Federation of American Scientists: http://nuclearweaponarchive.org.

35. My mentor, Luis Alvarez, was one who believed this. He designed the triggering electronics for the Nagasaki bomb. He also had the task of measuring the yield of the Hiroshima blast, so he was present in a second plane over Hiroshima when that city was destroyed. Alvarez attributed the German failure to the fact that their program was organized in a hierarchical manner, without the wide sharing of data that Oppenheimer insisted on in Los Alamos, so mistakes made by the top people (such as Heisenberg) were unlikely to be caught.

36. The excess radiation is 0.06 rem per year above the normal 0.2 rem per year. For 50,000 people for 50 years, that is a dose of 150,000 rem. Assuming the linear hypothesis, every 2500 rem leads to one death, so 150,000 rem should lead to $^{150,000}/_{2500}$ = 60 deaths.

37. This number is based on the fact that granite typically contains 4 parts per million of uranium. I take the dimensions of the Colorado Rockies watershed to be 300 by 400 kilometers, and I consider only rock from the surface to a depth of 1000 meters.

38. The dangers of plutonium are analyzed in detail in a report available at www.llnl.gov/csts/publications/sutcliffe/118825.html.

39. A clip of Cronkite saying this is shown in the *Frontline* program "Nuclear Reactions."

40. An electron won't work for this purpose because it is too light, and that makes its wave function spread out too much to neutralize the charge of the proton.

IV · SPACE

1. For the experts, the path is actually closer to an ellipse. That's because you are in orbit about a nearly spherically symmetrical mass, the Earth. A parabola is the limit for a flat Earth. For the superexperts, it isn't exactly an ellipse either, since the Earth is not totally spherically symmetrical.

2. At an altitude of 22,000 miles, the satellite is 26,000 miles from the center of the Earth. That means it is 6.5 times farther from the center than it would be if it were on the Earth's surface. If you've studied high school physics, then you know that Newton's Law of Gravity states that the force

of gravity decreases as the square of the distance. That makes the force of gravity $(6.5)^2 = 42$ times weaker than at the Earth's surface.

3. Suppose each satellite can photograph about $200 \times 2000 = 40,000$ square miles. Since the area of the Earth is about 200,000,000 square miles, it would take 5000 satellites for continuous and complete coverage.

4. The external tank holds 751 tons of fuel, and each of the two solid rocket boosters weighs 590 tons, for a total of 1931 tons.

5. The energy required is equal to the potential energy difference between the Earth's surface and infinity. Per gram, that energy is given by the formula $E = gR$, where R is the radius of the Earth. Putting in 980 cm/s^2 for g and 6.4×10^8 cm for R gives $E = 4.2 \times 10^{11}$ ergs = 15 food calories. That's the energy in 1.5 grams of gasoline.

6. The tall-building approach was once tried but led to a dismal failure, at least according to the biblical account of the Tower of Babel. The problem there, however, was not physics but the opposition of God. To confound the builders, God made the workers speak in mutually incomprehensible languages; this story gives rise to the verb babble.

7. Neutrinos do not "feel" either the electric force or the strong force. Photons don't feel the weak force.

8. To be more exact, there are 2.205 pounds in 1 kilogram, and 0.4536 kilogram in 1 pound, but don't bother memorizing these more precise numbers.

9. Most introductory physics textbooks insist on using different units for mass and weight. In their convention, a kilogram is always a mass, and a pound is always a weight. That is fine, but it doesn't help students understand why Europeans weigh food in kilograms. Keeping the units separate is useful mainly for people who are learning how to plug numbers into standard formulas.

10. Newton's Law of Gravity is usually written like this: $F = GMm/r^2$. G is a constant, M is the mass of the Earth, m is the mass of the object (for example, you), and r is the distance from the object to the center of the Earth. If you are on the surface of the Earth, so r is the Earth's radius, then the combination GMm/r^2 is given the name g. This is the same g that we use in the text. It is often called the acceleration of gravity, but it is that only at the Earth's surface.

11. Newton's Second Law is usually written $F = ma$, where F is the force in newtons, m is the mass in kilograms, and a is the acceleration in meters per second per second (m/s^2). This is not of much use to people who are not familiar with newtons or who are confused by square seconds. This equa-

tion can be written as $F = (mg)(a/g) = wA$. Here the force F and the weight w are in any force units you like; pounds work fine. $A = a/g$ is the acceleration measured in g's. This is the way I advocate teaching the equation to future presidents.

12. For the experts, this is how the calculation is done: the distance that you travel at acceleration g is given by the distance equation $D = \frac{1}{2}gt^2$. I'll do the calculation using metric units, in which $g = 9.8$ meters per second every second. We can calculate the number of seconds in a year as follows: 1 year $= 365$ days $= 365 \times 24$ hours $= 365 \times 24 \times 60$ minutes $= 365 \times 24 \times 60 \times 60$ seconds $= 3.16 \times 10^7$ seconds. Plugging in this value, we finally get $D = \frac{1}{2} \times 9.8 \times (3.16 \times 10^7)^2 = 5 \times 10^{15}$ meters. One light-year is the speed of light (3×10^8 meters per second) multiplied by the number of seconds in a year (which we just showed was 3.16×10^7). The result is 9.5×10^{15} meters. To keep the calculation simple, I ignored the space contraction and time dilation effects of special relativity.

13. Gravity follows an inverse square law, so from the small radius you would expect the force to be $(3.7)^2 = 13.7$ times larger. Combine that with the $\frac{1}{81}$ effect from the mass, and you get a net factor of $13.7/81 = \frac{1}{6}$.

14. For simplicity, assume that the asteroid has the same density as the Earth but is 1 mile in radius. That is 4200 times smaller than the radius of the Earth, making the mass $(4200)^3$ times less. But the small radius makes the gravity $(4200)^2$ times larger. Combining those factors reveals that the gravity is reduced by 4200. This answer is approximate, since the density of the asteroid might be different from that of the Earth.

15. Using physics, we can calculate the speed required to reach 60 miles (100 kilometers). I'll use metric units. The energy to reach height h is $E = mgh$, where g is the acceleration of gravity, approximately equal to 10 m/s². If $m = 1$ kilogram and $h = 60$ miles $= 100$ kilometers $= 1 \times 10^5$ meters, then $E = 1 \times 10 \times 10^5 = 10^6$ joules. The velocity that you need can be calculated from $E = \frac{1}{2}mv^2 = 1,000,000$ joules. Solving for v gives $v = \sqrt{2 \times 10^6} = 1414$ m/s $= 1.4$ km/s, approximately 1 mile per second.

16. The kinetic energy equation in physics is $E = \frac{1}{2}mv^2$. That equation requires a consistent set of units, which I will use here. For $m = 1$ kilogram and $v = 8000$ meters per second (the same as the orbit velocity of 5 miles per second), the equation gives $E = 32$ million joules $= 7650$ food calories, equal to 7.65 food calories per gram. TNT typically releases about 0.6 food calories per gram.

17. At Mach 1, the energy per molecule is roughly equal to the ordinary thermal energy. (That's because room-temperature molecules move at the

speed of sound.) At Mach 18, the energy per molecule is $18 \times 18 = 324$ times greater. That raises the temperature from 250 Kelvin (the temperature of the atmosphere at altitude) to 324×250 Kelvin $= 81,000$ Kelvin $= 146,000°F$.

18. The most dangerous job in the United States is the presidency. Four of the 42 presidents have been assassinated in office: Lincoln, Garfield, McKinley, and Kennedy. Unsuccessful attempts were made on Jackson, Truman, Ford, and Reagan. You should consider this fact before you run for the office.

19. After witnessing these events, I asked one border patrolman what he thought of those trying to sneak into the United States. "They are really decent people," he said. "They come here, work hard, and send most of their earnings back to their families in Mexico. I wish I could get my son to work that hard." The life of the border patrolman struck me as being similar to that of Sisyphus. And, as Albert Camus argued in his essay *The Myth of Sisyphus* Sisyphus was happy. See http://muller.lbl.gov/teaching/Physics10/pages/sisyphus.html.

20. Two thousand food calories per day is 2000×4200 joules per day $= 8.4 \times 10^6$ joules per day. One day consists of 24 hours of 60 minutes, each with 60 seconds, so the number of seconds in a day is $24 \times 60 \times 60 = 86,400$ seconds. Thus the number of joules per second is $8.4 \times 10^6/86,400 = 97$ watts—approximately 100 watts.

21. The army has adopted this motto too, because they now train in the same way for night combat.

22. In addition to the two IR systems, they use a night vision scope that, instead of IR, detects the minute visible light from stars. These are called *starlight scopes*.

23. About 10 microns, 40% of the diameter of a 25 micron fine human hair.

24. The wavelength is the speed of light divided by the frequency. Short wavelength means high frequency; long wavelength means low frequency. Near IR has a wavelength between 0.65 and 2 microns. Visible light has a wavelength near 0.5 micron.

V · GLOBAL WARMING

1. In the documentary, Gore attributes a version of this quote to Mark Twain. But it ain't so. Though it has a long record of being misattributed to Twain, the quote actually should be credited to Josh Billings, a nineteenth-century humorist. I checked with the Twain project at the University of California at Berkeley on this, and with a descendant of Billings.

2. In *An Inconvenient Truth*, Gore gives a superb example of how unreliable and untrustworthy a consensus can be. He tells the story of the strong consensus of the scientific community (consisting of the experts in this field) that South America did not once nestle up against Africa and then later separate from Africa to become its own continent. That was the derided idea of "continental drift." We now know that the consensus was wrong, and that the two continents indeed were once one. Continental drift, now called *plate tectonics*, has become a core concept in geology and geophysics. It is ironic that later in his movie, Gore admonishes us to accept the IPCC consensus.

3. The report is "Surface Temperature Reconstructions for the Last 2,000 Years" (National Academies Press, 2007). Reviewers for such reports are not anonymous, but are listed in the text. In addition to my role in this report, I have coauthored many papers and a technical book on the subject of climate history: *Ice Ages and Astronomical Causes* (Springer, 2000).

4. Barbara Tuchman, *A Distant Mirror* (New York: Knopf, 1978), 24.

5. In its 2007 report the IPCC states that it is "very likely" that humans are causing at least part of the warming "of the past 50 years." It defines *very likely* as 90% confidence. That is equivalent to saying that there is a 10% chance that none of the global warming is caused by humans. I'll discuss this in more detail in the chapter titled "A Very Likely Cause."

6. In *An Inconvenient Truth*, Al Gore shows this plot, removes it, and then states that the warmest year on record was 2006. He is not making a mistake; he is just choosing to refer to a different analysis—one produced by NASA scientist James Hansen, and not the one used by the consensus report. Hansen believes that his analysis is better than that of the IPCC.

7. The tests *were* responsible for increased radioactivity in the atmosphere, so Pauling's Nobel was well deserved.

8. The issue is a little more complicated than I suggest here. Clouds reflect sunlight during the day but can keep you warm at night by emitting IR. Moreover, some high-altitude clouds actually warm the surface during the day. But the net effect of clouds is to cool.

9. Most of the values on this plot were measured by extracting atmospheric gas from old glaciers in Greenland and Antarctica. The data points from 1958 to the present were made by direct measurement of the atmosphere by Charles Keeling, acting on a suggestion of Roger Revelle.

10. Because coal plants aren't completely efficient, 1 gigawatt of electricity requires 3 gigawatts of heat. For the experts, that's 3×10^9 joules per second, about 7×10^5 food calories per second. Each gram of coal has 6 food

calories, so the power plant consumes about 10^5 grams per second, or about 1 ton (a million grams) every 10 seconds.

11. In the 1960s, I sometimes used Freon-11 to wash my hands. It has the chemical formula $CFCl_3$, meaning that each molecule has one atom of carbon, one of fluorine, and three of chlorine.

12. Intergovernmental Panel on Climate Change, "Climate Change 2007: The Physical Science Basis: Summary for Policymakers" (Geneva, Switzerland: IPCC Secretariat, 2007), 10. The quote I give is taken directly from the document. In other parts of the report, however, a similar statement is made with the word *most* added, implying that there is a 90% chance that humans are responsible for *most* of the warming. The IPCC report doesn't explain this discrepancy in words. If we include the word *most*, then its conclusion implies that there is a 10% chance that *most* of the observed warming is due to something other than humans. These small changes in meaning are not important for the remainder of the discussion in this book.

13. Renyi Zhang et al., "Intensification of Pacific storm track linked to Asian pollution," *Proceedings of the National Academy of Sciences of the USA*, 104 (March 27, 2007): 5295–5299.

14. Al Gore does show the plot in his film *An Inconvenient Truth*.

15. Figure based on the 12th edition (2007) of the *Index of Leading Environmental Indicators*, published by the American Enterprise Institute for Public Policy Research (www.pacificresearch.org).

16. P. J. Webster, G. J. Holland, J. A. Curry, and H.-R. Chang, "Changes in Tropical Cyclone Number, Duration, and Intensity in a Warming Environment," *Science* (September 16, 2005): 1844–1846 (www.sciencemag.org).

17. The authors state in their report, "This trend is not inconsistent with recent climate model simulations that a doubling of CO_2 may increase the frequency of the most intense cyclones, although attribution of the 30-year trend to global warming would require a longer global data record and, especially, a deeper understanding of the role of hurricanes in the general circulation of the atmosphere and ocean, even in the present climate state."

18. Christopher W. Landsea, "Counting Atlantic Tropical Cyclones Back to 1900," *Eos* 88 (May 1, 2007), 197–200.

19. Figure based on the 12th edition (2007) of the *Index of Leading Environmental Indicators*, published by the American Enterprise Institute for Public Policy Research (www.pacificresearch.org).

20. The latest report is available at www.ncdc.noaa.gov/oa/climate/research/ monitoring.html.
21. Nicholas Caillon et al., "Timing of Atmospheric CO_2 and Antarctic Temperature Changes Across Termination III," Science 299 (March 14, 2003), 1728–1731.
22. I am rounding the numbers to make them easier to remember. More precise numbers are as follows: gasoline has a density of 0.8 gram per cubic centimeter and energy of 10 kilocalories per gram. Liquid hydrogen has a density of 0.068 gram per cubic centimeter and energy of 26 kilocalories per gram.
23. Technically it was too large to be a bomb, so historians refer to it as a hydrogen *device*.
24. The carbon dioxide emissions of China are about equal to ours, but China's population is 1.3 billion, compared to our 300 million.
25. Here's the math: Each year the efficiency improves by 1.02. To compound that for 55 years, calculate $(1.02)^{55} = 3$.
26. Some, notably economist Julian Simon, argue that the end of population growth will actually be bad for the economy.
27. Wolfgang Lutz, Warren Sanderson, and Sergei Scherbov, "The End of World Population Growth," Nature 412 (August 2, 2001), 543–545.
28. The paradox gets its name from the use in England of the commons, shared fields for grazing.
29. Less notable was the author of this book, who resigned from the Sierra Club in the early 1980s when they chose to oppose nuclear power. In my letter of resignation I cited the danger of greenhouse warming from the only feasible energy source alternative, fossil fuel.

INDEX